Andreas Koch, Peter Mandl (Eds.)

Modeling and Simulating Urban Processes

GEOSIMULATION

herausgegeben von

Univ.-Prof. Dr. Andreas Koch

(Universität Salzburg)

und

Ass.-Prof. Dr. Peter Mandl

(Universität Klagenfurt)

Band 1

LIT

Andreas Koch, Peter Mandl (Eds.)

Modeling and Simulating Urban Processes

LIT

Cover Picture: M. Seeger, 2006

This publication has been funded with support from the Austrian Federal Ministry for Science and Research (BMWF) and the "Stiftungs- und Förderungsgesellschaft der Paris-Lodron-Universität Salzburg"

UNIVERSITÄT SALZBURG

Bibliographic information published by the Deutsche Nationalbibliothek
The Deutsche Nationalbibliothek lists this publication in the Deutsche Nationalbibliografie; detailed bibliographic data are available in the Internet at http://dnb.d-nb.de.

ISBN 978-3-643-50036-6

A catalogue record for this book is available from the British Library

Krotenthallergasse 10/8
A-1080 Wien
Tel. +43 (0) 1-409 56 61
Fax +43 (0) 1-409 56 97
e-Mail: wien@lit-verlag.at
http://www.lit-verlag.at

LIT VERLAG Dr. W. Hopf
Berlin 2011
Fresnostr. 2
D-48159 Münster
Tel. +49 (0) 2 51-620 320
Fax +49 (0) 2 51-922 60 99
e-Mail: lit@lit-verlag.de
http://www.lit-verlag.de

Distribution:

In Germany: LIT Verlag Fresnostr. 2, D-48159 Münster
Tel. +49 (0) 2 51-620 32 22, Fax +49 (0) 2 51-922 60 99, e-mail: vertrieb@lit-verlag.de

In Austria: Medienlogistik Pichler-ÖBZ, e-mail: mlo@medien-logistik.at

In the UK: Global Book Marketing, e-mail: mo@centralbooks.com

In North America by:

Transaction Publishers
New Brunswick (U.S.A.) and London (U.K.)

Transaction Publishers
Rutgers University
35 Berrue Circle
Piscataway, NJ 08854

Phone: +1 (732) 445 - 2280
Fax: + 1 (732) 445 - 3138
for orders (U. S. only):
toll free (888) 999 - 6778
e-mail: orders@transactionpub.com

Contents

Preface

The investigation of change – of climate, urban or social structures, technological or cultural innovations – and the related effects, e.g. on biodiversity, migration, segregation or commu-nication, are one of the core issues in geography. Different spatial and temporal scales are crucial to analyze and interpret changes of our environment. Moreover, it is possible to dis-tinguish retrospective and prognostic techniques, local and global, dynamic or linear devel-opments. All these phenomena point out that change is commonly understood as a complex research objective, taking multiple relationships into account. Complex processes can range from direct cause-effect chains to alterations of parameter values or evolutionary develop-ments.

Modeling and (geo-)simulation techniques offer a wide range of opportunities which cannot completely or adequately be accomplished by traditional quantitative or qualitative methods. Depending on the model purpose, different approaches like system dynamics (studying the behavior of an entire model), cellular-automata or agent-based modeling (studying the beha-vior of model elements and relations), or network analysis (studying the structure of relation-ships) can be applied. By visualizing and explicitly representing different scales and relations, we are able to intuitively better understand spatio-temporal processes.

In addition, simulation models suitably represent some kind of laboratory. Processes which proceed very fast or very slow in reality can be stretched or squashed artificially in such labs. In doing so, we get new insights in yet hidden relationships. Furthermore, scenarios can be generated without damaging the natural and social environment and without intervening in a morally reprehensible way. The capability to visualize the simulated processes comprehen-sively supports collaborative work.

All these topics are strong arguments for the modeling and – here – geosimulation domain. The idea to release a series about such topics dates back to the year 2003 when the editors of this volume had organized a session about multi-agent-systems in geography at the German-speaking geography conference in Bern (Switzerland). Four years later, a second session titled modeling and simulating complex geographical systems followed in Bayreuth (Germany). Now, with the third session of that type which took place in Vienna (Austria) in 2009, the editors decided to launch a new series that captures all of the interesting issues around the notion of geosimulation. Volume 1 is specifically dedicated to urban processes.

Andreas Koch reflects on the theoretical, epistemological and methodological conditions to link geosimulation modeling techniques with social geographical themes. The theoretical challenge is recognized in coupling the social worlds with the spatial worlds which are as-sumed to be two independent spheres with respective properties. Social and virtual spaces, as two instances of this coupling process, are discussed in detail. In doing so, a system theo-retical approach is proposed which enables to focus on both the analytical level (with the social and spatial entities) and the synoptical level (the relationships between these entities). With this theoretical frame, Koch argues, it is possible to represent social-temporal-spatial scaling at different levels explicitly. Geosimulation with its diverse methodological approaches – be it a Cellular Automata model (CA) or an Agent-Based Model (ABM) – is a promising tool and method to meet these theoretical needs. By referring to complexity theory and its associated concepts like emergence, interaction or iteration, the article illustrates the coup-ling of theory and method. The final coupling between theory and method on one hand and epistemology on the other is done by a discussion about the supposed gap between positivist techniques of CA, ABM, or Geographic Information Science (GIS) and constructivist critically inspired social geography. Koch concludes that neither simple positivism nor radical constructivism can be an acceptable solution, but a mutually reference of the two, accepting construction and re-construction, production and re-production with the latter unfolding some kind of pre-given facts.

Alexandra Hill and Christian Lindner choose a Cellular-Automaton-based (CA) model ap-proach to describe and explain the spatial impact of informal urban growth in Dar es Salaam, Tanzania. The growth of informal settlements is still a major planning issue in many large cities of less developed countries, and geosimulation appears to be a promising tool and me-thod to reveal the complex processes and relationships of residential migration patterns. Hill and Lindner propose a multivariate logistic regression model to achieve some explorative explanations with regard to settlement dynamics as well as infrastructural conditions that enforce informal growth. The aim of the CA-model which follows the regression model is to calculate the transition probability of each cell (approx. 100.000 altogether) whereby each cell represents a binary datum (vacant cell or informal residential cell). The model results are then compared with real-world development. In addition, transport infrastructure scenarios have been compiled in order to correlate spatial structure with spatial processes. The paper clearly points out that geosimulation is an appropriate tool to support planning decisions by making different levels of interaction relationships explicit and transparent.

A similar approach with respect to methodological and planning-theoretical claims is proposed by Rudolf Giffinger and Roman Seidl. They too advocate for modeling and simulation techniques as suitable instruments to attend profound changes that took place in the spatial planning domain. While Hill and Lindner focus primarily on the structural outcomes of micro-processes, Giffinger and Seidl do so on the processes itself – which is gentrification in their case. They raise the question which meaning micro-modeling has in planning processes theoretically and practically. The role of spatial planning, its understanding of influencing spatial development, as well as the role of their methods and tools applied changed over the last decade significantly. The article is addressed towards a model approach which promotes ideas of mutually linked supply and demand side theories of gentrification, including aspects of complexity theory. In doing so, a micro-simulation formulation is introduced. Against the background of one-sided theories of gentrification like rent-gap-, hedonic-price-, or Alonso-Muth-approaches, emphasizing economic or lifestyle issues, they prefer a specification of the double invasion-succession-cycle theory by including the idea of tipping-point process. Com-plex-theoretical aspects like self-organized criticality are implemented due linking interactions between spatial and social entities. Tolerance of social agents, a critical topic in segregation research, which is defined here as accepted shares of different groups of social agents within a certain neighborhood range, complexifies neighborhood dynamics in gentrification processes. Tolerance does not only depend on intra-group, but also on inter-group conditions. The proposed model is mainly theory-driven and thus represents a kind of laboratory environment that still has to be linked to empirical data. Finally they discuss the problem of communicating such models to planning experts unfamiliar with these methods.

The concept of tolerance is also central in the contribution of Reinhard Koenig. He first intro-duces the remarkable thesis that the size of an urban settlement has an essential effect on the socio-spatial organization and the residential segregation of its population. To proof this thesis, urban structure – from streets and buildings to dwellings – is represented as a graph. While nodes represent street segments, buildings and dwellings, edges represent the spatial connections among these entities. This kind of modeled city which Koenig denotes as circle city due to its respective shape enables the investigation of every possible spatial configura-tion. In order to study the graph's properties comprehensively an agent-based segregation model is applied by him. The model analysis results illustrate that there are relevant effects of the size of urban settlements on the socio-spatial organization of its population if the tolerance threshold is within a critical range. This range refers to an interval within which a small change causes a relatively large outcome on the segregation of urban population. Moreover, a positive correlation between the size of a city and the segregation of the population can be shown. This relationship increases in intensity if the density of the settlement increases. A higher density can be represented in the model, for example with more houses per street or more dwellings per house.

With a simulation model of urban land-use change of the German state of North Rhine-Westphalia we come back to the Cellular Automata approach. Roland Goetzke and Michael Judex introduce a comprehensive modelling framework called XULU that integrates impor-tant functions like visualization and model control. XULU thus serves as a software-independent platform (Java, however, is needed), and in this article it hosts two land-use models with different functions. Goetzke and Judex compare the potentials and bottlenecks of Clue-S, a statistical package to analyze land-use change, and SLEUTH, a CA-based urban growth and land-use change model. They conclude that there is no general ranking, but a task-specific consideration necessary. If an explicit focus on urban growth is given then programs like SLEUTH should be preferred; if a more general approach to land-use changes, including rural areas, is pursued then a statistical package like Clue-S has some advantages. Again, the methodological approach is seen appropriate in dealing with planning issues, because simulation models appear to have the potential to assist decision makers in analyzing underlying (sometimes hidden) processes and assessing relational impacts. One crucial premise, however, is that model calibration and validation actually takes place.

Christina West and Philipp Deschermeier pose the question: "Who is living where and why?". With this question they imply an individual value-oriented approach when studying processes of intra-urban and suburban residential migration. However, the micro-perspective in their contribution is methodologically translated by applying a spatial econometrics approach. Based on empirical data about value-orientation and residential satisfaction of inhabitants of the German city of Mannheim, they first adopt multivariate statistics (factor and cluster analy-sis) to classify urban space into different social districts. Due to the assumed fact that local satisfaction correlates with environmental conditions that includes neighbored districts too, they secondly analyze spatial autocorrelation and spatial heterogeneity with techniques of spatial econometrics. West and Deschermeier understand satisfaction and value, respectively as being distance-sensitive and thus applicable to econometric modeling. Being aware of statistical restrictions caused by a pre-given low spatial sample size, they conclude that for some cluster types spatial autocorrelation – and thus a perspective beyond the local view – is indeed an influencing issue if residential satisfaction comes into play.

The editors would like to thank all authors for their contributions to this volume. Furthermore, we appreciate it very much that LIT publisher demonstrated great interest in our engagement. They decided to launch this series and they supported us in proof-reading for which we would like to thank very much. We also would like to thank the Salzburger Stiftungs- und Förderungsgesellschaft and the Austrian Federal Ministry for Education, Arts and Culture for their support to publish this volume.

Andreas Koch & Peter Mandl

Salzburg und Klagenfurt, August 2010

Geosimulation: theoretical and epistemological expeditions to the invisible relationships of space, time and social life

Andreas Koch

Introduction

The letter ***S*** in GIS has altered its meaning several times over the past two decades or so. One effect of this process is a sequential overlay of content rather than a sequential succession. During the infancy of the emerging field of geocomputation, it was the notion of geographic information *systems* which was (and still is, of course) associated with a technological translation of spatial data and their coupling with non-spatial attributes (Bartleme 2005; de Lange 2005). Subsequent progress in data acquisition (e.g. via GPS or remote sensing; mobile and real-time), handling (GeoDBMS), analysis (e.g. techniques of spatial analysis, geostatistics, network analysis or tracking analysis) and visualization (e.g. 3D city models) as well as data interoperability (e.g. SDI, OGC, INSPIRE and GMES) has increasingly raised questions about the potentials and limitations of methodological translations of geographical theories. In this context geographic information science became a more prominent translation of the acronym (Goodchild 1992; Longley 2010).

Scientific discourses seeing geographic information as a research objective rather than a research tool have been accompanied by the growing popularity of virtual multimedia systems such as Web 2.0, Google Earth and social networking sites. These have provided another translation, one emphasizing the everyday praxis of people; ***S*** can thus also represent GI*Society* (Fischer 2009; Strobl 2009). Against this background of scientific and other translations of geographic information, one could easily add another dimension; a GISimulation approach which captures the different spatial-temporal scales of processes. In this case a scaling translation would be applied.

With these diverse translations of the letter ***S*** a different but nonetheless specific meaning is associated in each case. The letter ***G***, however, seems to have a consistently less diverse meaning. This assumption justifies further exploring the question as to what the notion of geographic really stands for. In doing so, this article refers to the more commonly used term of geosimulation rather than GISimulation. The starting point of subsequent reflections about the theoretical, methodological and epistemological impact of space should be to tackle the question of why it is important to consider spatial relationships. In other words: what functions do spaces have when modeling and simulating geographic phenomena, what purposes do spaces serve, how are spaces imagined and how do we achieve adequate visualizations?

The above-mentioned dimensions of translation support a differentiated pro cedure for this subject matter.

Accordingly, the agenda is threefold. First, chapter 2 reviews some theoretical thoughts about space and spatiality and their representational modes. The focus here is specifically directed on virtual and social spaces. Second, chapter 4 introduces a theoretical synthesis of space, time and sociality, which can be appropriately applied to geosimulation models. This synthesis is, thirdly, completed with a methodological and epistemological consideration of the aforementioned triad (chapter 5). For illustration and explanatory purposes, chapter 3 presents an additional segregation model that attempts to meet the requirements outlined so far.

GEO-Simulation: thoughts about spatiality, its representation and visualization in agent-based geosimulation models

In his recent publication "Theories", the philosopher Martin Seel reflects on seeing the 'world' as a reality which is strongly related to the individual who consciously reflects subjectively, but in a way, however, which can also be inter-subjectively imagined as an intertwining and juxtaposing of states and processes (Seel 2009: 69). Reality understood in this way is not completely at an individual's disposal, but is so in a specific manner. This specification in turn correlates with, amongst others things, representations of a 'world' somebody has in mind. Accordingly, by focusing on space and spatiality instead of an abstract world, we can conclude that spatial representations refer to functions dedicated to mediating the gap between thought and communication on one hand and the external environment on the other.

Virtual spaces

This gap can be bridged in manifold ways. While for instance phenomenological representations of space refer primarily to qualitative properties expressed in texts or cognitive maps, agent-based geosi-mulation techniques refer primarily to quantitative and formalized aspects of spatial entities as compu-tational artifacts. In general, different imaginations of experienced spatialities lead to diverse geographical conceptualizations (and vice versa), which at the same time share some common characteristics due to the independent properties of the spatial sphere. With ever more advanced information and communications technologies changing an individual's options for acting and social potential to participate, virtual and hybrid spaces have been introduced just as chaos theory, complexity theory, cybernetics and mathematical topology did scientifically. Some researchers arguing the relevance of virtual spaces reflect about that type of space "[...] not as a standardized and homogeneous structure like the Euclidean and Cartesian age would refer to as, but as an instantaneous configuration within chaotic processes, as a dynamic network of transitions inducing continuously new possibilities of connections" (Doetsch 2006: 207; translation A.K.). A lebensraum (habitat) imagined with such different concepts of space fundamentally changes the very conditions of imagination, representation and visualization, because space and spatiality are perceived as profoundly abstract and invisible with regard to topological structure. It thus can be argued that the function or purpose of a geosimulation modeling approach is to represent virtual/hybrid spaces by making them visible and concrete.

This idea can be illustrated appropriately by Flusser's (1991) reflections about space. In his introduction, Doetsch (2006: 207f) points out that, with respect to Flusser, virtual space (like outer space) cannot be adequately derived by bodily experience or by language alone, but by applying mathematical algorithms. This approach can be seen as an attempt to overcome constructions of space that involve a priori assumptions and to offer different ways of generating new and diverse realities, thereby understanding space as an interdependent force. In addition, an important clue is expressed as to the contingencies of mutual translations between the social and the spatial spheres: while technological and methodological modes of spatial modeling seem to be more or less easily solvable, methods of scaling models and of integrating every-day perceptions appear to be a significant obstacle. This is due to the fact that relationships in topological space manifest simultaneously at a local and global level, as can be observed particularly in stock trading and market dynamics.

The significant transition that occurs when virtual space is thought of as inherently incorporated into Lebenswelt, is that space and time in general have to be imagined as being interwoven rather than being interrelated (see Flusser 1991: 275). The difference made here is that space and time are not assumed to be two distinct phenomena, but inextricable space-time. This is why we experience and witness physical-virtual lebensraum much more as a space in which events do not occur certainly but probably. One issue associated with this transition refers to the modified relationship occurring between public and private space. The traditional discrimination of public and private spaces which correlates with distinct territorial boundaries and the concept of geometrical space, has been replaced by an ongoing process of dissolution and amalgamation of the two: "The separation between public and private is less useful if so called politicians are able to arrive uninvited in the kitchen via cable transmission" (Flusser 1991: 280; translation A.K.).

The conceptual shift from territory to topology and the interweaving of space and time accordingly requires methodological procedures that are able to turn hardly visible spatial activities into imaginable representations, "[...] to introduce spaces filled with things (ideas, objects, etc.) which could not be thought of before" (Flusser 1991: 281; translation A.K.). Geosimulation techniques with their manifold possibilities of representing and visualizing space-time phenomena appear to be a suitable approach in this regard. Imaginations and ideas about virtual space bridge phenomenological space concepts which attempt to relate the diversity of subjective and contemporary experience to general causes, with social space concepts which adopt these general causes as a starting point for distinguished studies of spatial organization and order. Because social spaces are constituted through social patterns of structuring, their representation by geomodels is likely as challenging as it is for phenomenological and virtual spaces.

Social spaces

Sociology and social geography have not only developed the idea of a social space over the last century, but have also shifted the concept of space from one of natural determinism to a socially constituted spatiality. The spatial description of social organization is thus less concerned with the structural order of space, but rather the practice of acting which constitutes space and place (see Dünne 2006: 290). Foucault (1984) identified this transformation as a sequence of placing-extension-location. The dominant placing paradigm during the middle ages, most notably dedicated to distinguishing profane and sacred places, served an additional function of creating social order. Since this social order was established as a common goal, placing shifted to extension and accordingly to geometry (Dünne 2006: 293f). Progress in cartography in the 17th and 18th century is an example of this change. As both paradigms matured in social contexts and have been further developed by technological, cultural and economic changes simultaneously, confident patterns of social order and organization were also resolved or changed. On such a line postmodern societies are characterized by a relational constitution of socio-spatial order which promotes the social and temporal context of generating spatiality, instead of depending on a pre-determined setting of spatial meaning. Hence Foucault's reflections on contemporary spatial concepts refer to a similar relational dimension to that proposed by Flusser for his virtual space. Indeed, the formalization of 'imagining space relationally' is considered explicitly by Foucault (1984: 318): "Locations are determined by neighborhood relationships between points or entities that can be described formally as mathematical series, trees or grids". The dichotomy of public and private space seems to be an anachronism for him as well.

The three phases of spatial reasoning (in a non-deterministic sense) in Foucault's approach are in some ways orthogonal to Lefebvre's (1991) proposal, who claims a production of space. Orthogonal because all three phases are characterized by three types of spatial production which can be traced back to the Greek polis (Lefebvre 1974: 331).

"1 *Spatial practice*, which embraces production and reproduction, and the particular locations and spatial sets characteristic of each social formation. Spatial practice ensures continuity and some degree of cohesion. [...]
2 *Representations of space*, which are tied to the relations of production and to the 'order' which those relations impose, and hence to knowledge, to signs, to codes, and to 'frontal' relations.
3 *Representational spaces*, embodying complex symbolisms, sometimes coded, sometimes not, linked to the clandestine or underground side of social life, as also to art [...]" (Lefebvre 1991: 33).

Spatial practice (in neo-capitalism) articulates at the level of perceived space and ties everyday reality to urban reality. Perceived space rests upon experience and refers to the physical component of the production of space. Representations of space are cognitively-produced spaces which involve the role of knowledge production. This level is crucial for spatial modeling, because this is "[...] the space of the scientists, spatial planners, urbanites, technocrats who identify embedded and perceived space with conceptualized space [...]. [And this space, A.K.] is the dominant space of a society (a production re-gime)" (Lefebvre 1974: 336). In life, embedded space refers to at least the symbolic component of space production and thus, as representational spaces, also to the creation of meaning in a social context. Remarkably, Lefebvre's approach takes the complementary connection of production and product, of construction and reconstruction explicitly into account; they are not two separate representations, but are intrinsically tied together. However, he also points out that the technique of analysis alone is not sufficient to capture the very nature of space; theoretical reasoning as an ordinary activity is crucial too, otherwise it is not possible to produce space in such a synoptic way. In other words, space considered exclusively as an abstract model remains at a poor level of perception, conceptualization and symbolization (Lefebvre 1974: 338). This statement very clearly points out that geosimulation techniques do not – and could not – cover the full range of spatial and social representation, but to be fair this is something neither technique is able to do.

Finally we turn our attention to some of Cassirer's (1931) ideas of theoretical space which he inte-grates into concepts of aesthetical and mythical spaces. His spatial approach provides us with the 'tools' we need to connect all the fragments presented above and achieve a suitable frame for modeling and simulating spatial states, functions and processes as an explicit and interdependent part of social simulation, i.e. a move towards socio-spatial simulation. Cassirer argues that the notion of order is privileged against the notion of being (in an ontological sense). The reason for this not self-evident shift is that being – the social as well as the spatial being – is set as an absolute unity. Being is inherently a comprehensive and monistic totality, with an identity that can only develop if it denies itself (Cassirer 1931: 491). The notion of order, however, always refers to diversity and difference. Therefore Cassirer accordingly argues for a shift in thinking space, in order to achieve plurality rather than uniformity when investigating social phenomena. With respect to modern physics, he argues that space is no longer a thing among things. Things are neither in time nor in space. Rather the world represents a system of events, spatially and temporally interwoven (Cassirer 1931: 490). A shift from geometrical to topological concepts of space is implicitly included. This shift does not mean exclusion however, so geometry still has its meaning; but compared to former approaches, this meaning is relatively decreased. Moreover, this shift makes it obvious that it is important to incorporate space in geographical simulation models – as an attempt to create imaginable spaces from abstract concepts.

A brief conclusion to the task of modeling and simulation in (human) geography can be drawn here: Besides the interweaving of space and time it is the interweaving of spatial and social organization that has to be appropriately taken into account when translating social spaces into geosimulation models - although there is no objective reference for 'appropriate'. The aim is to transfer abstract concepts into comprehensive models without denying abstraction. Temporal and spatial scaling must also be neces-sarily integrated in order to adequately model neighborhood and closeness in their topological and geometrical forms. It is also important to develop a concise methodological translation, based on a sufficiently general theory which allows for a coherent implementation of space-time-society relationships. Mandl (2003: 17) offers a matrix of differentiated spatial-temporal scaling in geo-modeling and geosimulation that can be used as a proper reference for our intention of translating experienced worlds into model worlds.

In addition to this a theoretical system proposal is introduced now, in order to theoretically implement space-society interdependencies in geosimulation models. This approach is intended to balance the interdependencies by complementarily confronting the process of the social construction of space with the spatial construction of society. Beyond all the functions and objectives an explicit implementation of space can provide for the modeling of social phenomena, it has been clearly pointed out by the spatial theories presented above how space is involved in processes of social organization, change and order. Space considered as a priori, as premise, as construction or as product: without an interdependent spatial category every description and explanation of social structures and processes remains imperfect. This exact background is one of the core driving forces with which to appropriately represent and visualize space in geosimulation. Writing from a sociological perspective, Dünne (2006: 296) suggests system theory as one possible solution with which to refer space to social processes, which neither argues with structural faults (like Foucault did) nor with teleological as-sumptions (Marc Auge's non-places are an example in this context, because they represent the ongoing process of the dissolution of spatial and temporal ties); with system theory one can instead observe the creation of social and spatial systems as a result of functional relations, structurally coupled and mutually reinforcing their respective existences.

One brief final remark: The selection of concepts of space presented here are of course not comprehensive – but have been made consciously. On the one hand they are important concepts of human geographical thinking, while on the other they establish an explicit tie of the discussion of transferability to the modeling and simulation spheres (for a discussion of several approaches to social space concepts in social geography and their applicability compared with system theory, see Koch 2004).

Simulating social-spatial interactions: an example

In this chapter a segregation model is presented which represents a specific type of social-spatial interaction model. The intention is to illustrate a few aspects of the theoretical and epistemological reflections which will follow in the next two chapters. The model is an attempt to operationalize the mutual relationship between the social construction of space and the spatial construction of society at a local scale. In doing so we focus on two issues relevant to the theoretical discussion of both social and social-spatial interaction. Social interaction is modeled as a multivariate phenomenon occurring among social agents equipped with several socially-indicated properties which influence their behavior as a whole. Hence we extend Schelling's (1969, 1971) approach to more than one variable, without forgetting to appreciate his convincing results on emergence. Social interaction is also modeled as a phenomenon of similarity, which means that agents' decisions – in this case whether to migrate or not – are based on those social attributes of other agents in their neighborhood that are similar to their own, rather than identical. This multivariate and similarity approach offers a more realistic view of interaction modeling and simulation, since residential mobility is no easy decision to make. This, however, complexifies the scaling of processes of social homogeneity and, accordingly, the recognition of patterns. Social-spatial interaction at least is modeled as a procedure in which social agents are able to actively change the local-spatial composition. This in turn changes the local-social composition, due to the fact that the changing the whole range of infrastructure - although we are aware of its importance (implicitly at best, because we assume that infrastructural aspects are correlated in some way to the variables that are included, such as real estate price or house-hold structure). The model itself is briefly described (for more details see Koch 2009a, 2010), as we concentrate on the facts that are important in a theoretical-epistemological context. Figure 1 illustrates the model structure.

Figure 1: model basics and simulation procedure (source: Koch 2010)

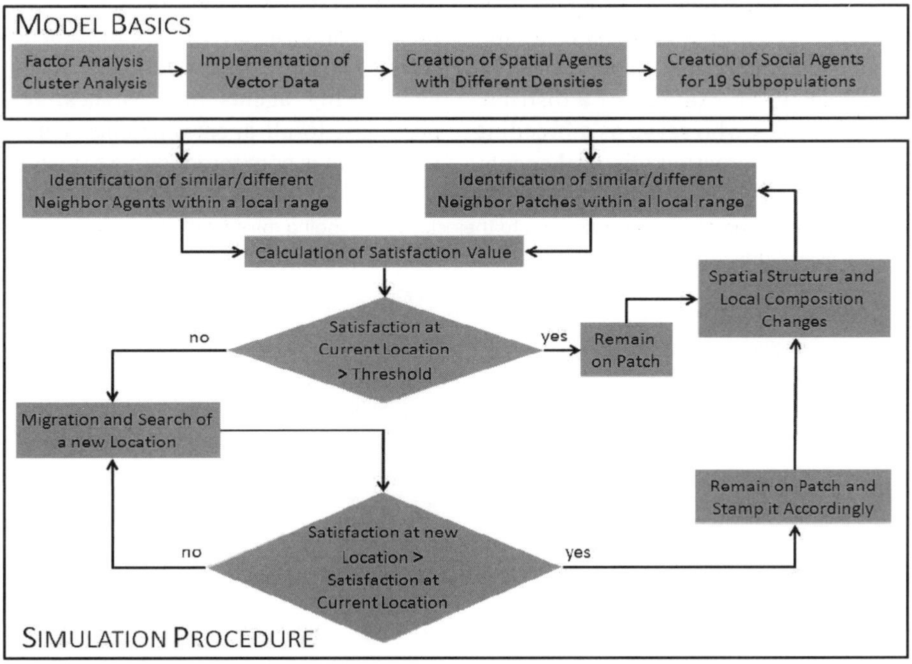

multivariate set of socialized spatial attributes enforces migration if dissimilarity exceeds a specific threshold.

Segregation in this context refers to social segregation and is thus not restricted to ethnic or any other single attribute form of segregation. The driving force behind this is residential mobility caused by per-ception of the social environment; we can therefore exclude

The model uses census data of the city of Salzburg, Austria, at the level of statistical districts. In order to derive social and spatial attributes, factor analysis and cluster analysis have been applied. Four factors have been discriminated, representing (1) socio-economic structure, (2) socio-demographic structure, (3) real estate market and (4) social status. The results of these analyses have then been disaggregated to the so-called patch level, a cell in the simulation program NetLogo (Wilensky 1999) representing a discrete spatial unit (patch, cell and spatial agent are used synonymously in this article). The initial artificial distribution – each statistical district is in fact a totally homogeneous area representing one of 19 cluster types – has been shuffled in the next step, taking claims of similarity into account. Each patch then, in general, serves as a 'birth place' for the creation of a social agent who inherits the social attributes of that patch. In order to keep the simulation model executable, a sample of social agents has been drawn. Figure 2 shows an example of a distribution of patches who serve as a breeding ground for the creation of social agents.

The evaluation of the social-spatial neighborhood is achieved by local knowledge, gained by interaction among social agents and between social agents and spatial agents. Knowledge of the locations visited due to migration depends only on spatial information; this can be explained pragmatically since potential new 'inhabitants' tend to gather information about only the spatial environment, much less about the social environment. Local social knowledge is calculated as a satisfaction value whereby local scaling can vary with user interaction. The calculated satisfaction value at an agent's present location (expressed as a ratio of similar neighboring agents and patches, respectively, to all agents/ patches within this range) is compared with a threshold

Figure 2: distribution of patches according to their cluster belonging mixed with other cluster types (source: Koch 2010)

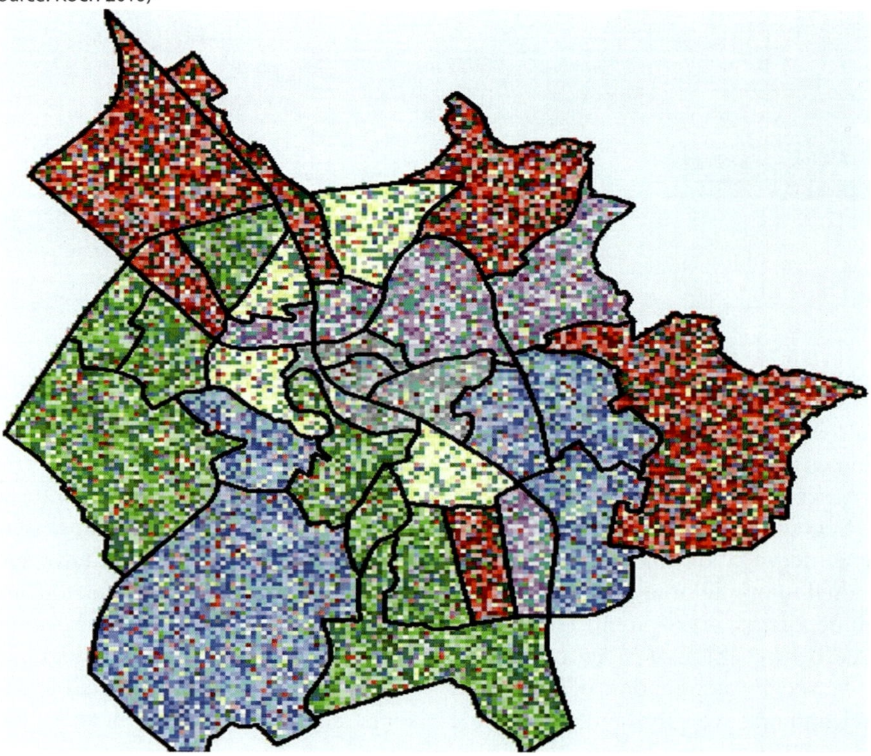

value in the next step. If the satisfaction value is below this threshold, then agents decide to move; otherwise they stay at their current location for at least as long as the heterogeneous spatial composition remains more or less unchanged by other agents, i.e. social agents are forced to migrate due to decreasing attractiveness of a place. This implemented process is intended to simulate the interaction between social and spatial systems (see Figure 3).

Figure 3: interaction pattern of social and spatial agents. Left: initial situation. Right: a red agent occupies a green patch and stamps it accordingly; a blue agent might be enforced now to migrate (source: Koch 2010)

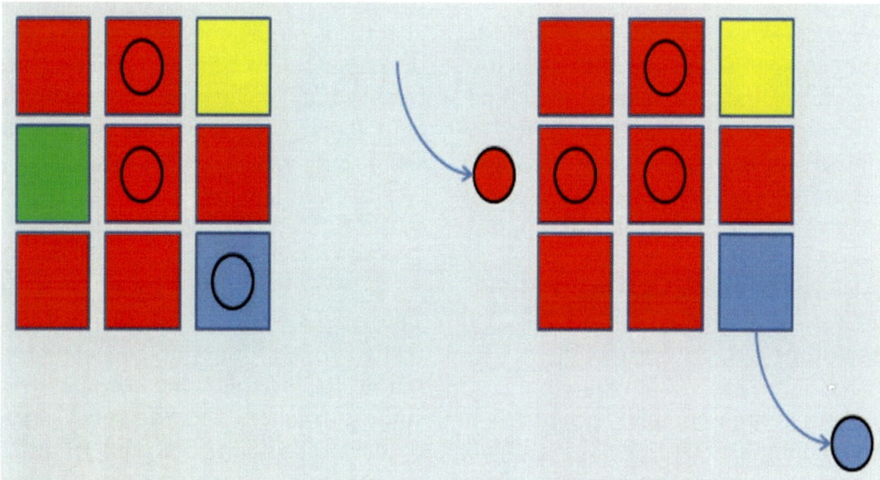

The dynamic is produced by the mutual correlation between agents' decisions and patch composition. Technologically the procedure is realized as changes in patch color and thus is directly tied to the second claim, that of similarity.

As mentioned above, the notion of similarity refers to the ability of social agents to differentially accept other agents of different social types (expressed as cluster types). The satisfaction value thus includes agents/patches of the same type and of similar types. Differences in similarity result from the clustering procedure (see Figure 4) and from the programmed ability of agents to accept other agents of the same type and of one level higher or lower with respect to household size, income or social status.

If the spatial environment is sufficiently heterogeneous at the model's initialization step, one can then conclude that it is possible to adequately meet some of the theoretical and epistemological claims that will be discussed in

Figure 4: clustering of the statistical districts of the city of Salzburg according to four factors

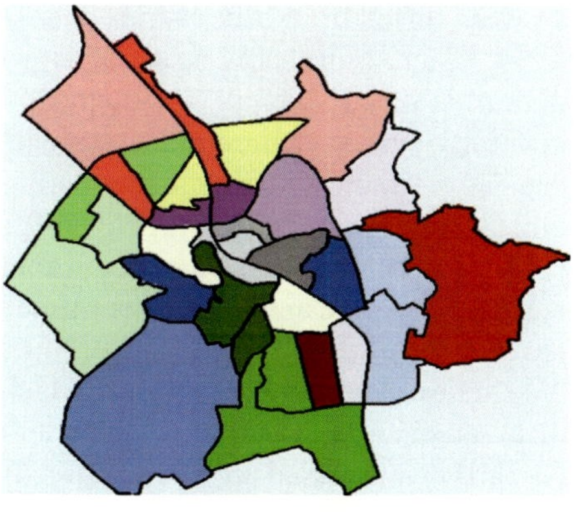

the next chapters. As I have stated elsewhere (Koch 2009a; with regard to modeling complex systems theoretically), "[...] we can say that systemic complexity is realized by the selective interaction behavior among agents. With regard to interactions between agents and patches, we observe that agents, by stamping their attributes at new locations, influence the spatial and, consequently, the social configuration of neighbourhoods in the district. Simultaneously, they are influenced by such actions of other agents. Systemic complexity, understood as reducing environmental complexity to "survive" socially can result in emerging patterns of social-spatial homogeneity which in turn means that complex social and spatial systems arise". The aim of the subsequent two chapters is to illustrate theoretical and epistemological questions against this explanatory background.

System and complexity theory: theoretical synthesis of a space-time-society approach

Reflecting upon complex systems is in itself a complex task since it is difficult to precisely define phenomena which are referred to as „complex". Accordingly, definitions often end up by merely listing characteristics in which complex systems differ from other types of systems. "There is probably no satisfactory definition of complexity, although it is generally agreed that complex systems consist of "many basic but interacting units" (Coveney and Highfield 1995, p. 7) and are "systems in process that constantly evolve and unfold over time" (Arthur 1999, p. 107)" as Batty (2005: 63f) states. Another reason for the lack of a satisfactory definition which will be further discussed in this chapter is the doubling of complexity in system theory approaches.

Luhmann's system theory (1993) claims that the reason for the existence of systems lies in the necessity to reduce the complexity of the world in order to enable communication. According to this theory, every system is less complex than its corresponding environment – yet it is nonetheless complex! This kind of complexity, here referred to as „systemic complexity", is therefore concerned with its internal systemic context. Whereas this point of view is one main aspect of research in the field of complex systems, one can also focus on the connections and interdependences of various different systems which mean that there are two complementary scientific approaches on the subject of complex systems.

Systemic Complexity - Complex Systems

The general statement that systems really do exist must be considered essential for the following considerations since it renders the theoretical basis of these considerations and its implicit consequences transparent. Furthermore, this general differentiation leads to further differentiations concerning, among other aspects, the type of system that is the distinction between social and spatial systems. Thus, the question arises in which ways these different kinds of systems are mutually coupled. From an epistemological perspective, this is primarily part of a general scientific approach which rejects any kind of subjective-arbitrary definition and distinction of systems focusing mainly on research issues. We see systems as epistemic objects, in other words, things "that are representations of concepts"

(Rheinberger 2001: 15). Therefore, epistemic objects serve as reference points for scientific work and experimenting.

One of the main challenges of modeling and simulating systems is the visualization of the non-visible, yet nevertheless cognitively imaginable. This, however, involves working with the representation of a system and not the actual system itself. It is therefore of utmost importance to distinguish between representation and the object itself. Nevertheless, simulation theory with its concept of „pragmatic construction of reality" assumes that a translation from representation into object can be successfully accomplished. Küppers et al. (2006: 8) explicitly state that computer simulations are "observational instruments": They [simulations; A.K.] can enable access to complex patterns of behavior undetected by classical instruments such as telescopes or microscopes. Whereas telescopes and microscopes render phenomena visible by affecting the scale of 'tangible' entities through optical processes of resolution, simulation renders 'visible' the effects of parameters and forces such as time, dynamic interactions, and so forth [...]. Thus, simulation, by constructing images, may translate absolutely non-visual events into a visual media!".

The following considerations are mainly based on Luhmann's theory of social systems and its translation into a theory of spatial systems (Koch 2004). This approach differs to a great extent from other studies which refer to system theory for the interpretation of spatial phenomena. To give a few examples, Klüter (1986, 1999) regards "space as an element of social communication", Wirth (1979) claims that spatial systems are system models, Stichweh (1998, 2003) interprets space as a medium and system environment and Fliedner (1999, 2001a, 2001b) views space as a scale-dependant type of system. However, we see here a significant affinity between the approach of this study and the social-ecological theory of Fischer-Kowalski and Erb (2006: 37).

One can consider the most important [accomplishment of system theory, A.K.] that it does not regard a system as a composition of various elements but as an operatively closed (autopoietic) process of self-reproduction which is separated from its environment by certain borders. On the other hand, this conception of systems allows for the possibility of quantifications and is quite compatible with various other modeling approaches [!]" (translation A.K.).

Systemic complexity

Systemic complexity describes the necessary process of reducing system's complexity in relation to its environment – each system generally has less complexity than its environment. Why? „The concept of complexity describes the fact that not all elements of an entity can be simultaneously interconnected. Therefore, complexity means that a certain selection is necessary to update relations between elements" (Baraldi et al. 1997: 93). The reduction of complexity is a necessary prerequisite for maintaining systems and for giving them meaning. The environment itself is so complex that selection and limitation to certain aspects is necessary for a system to relate to it meaningfully. This represents one side of systemic complexity. Its complementary counterpart is contingency, because there are continually more possibilities for connecting communication and thoughts than can actually be realized.

This dialectic leads to a reduction of complexity and simultaneously to an emergence of complexity (Krieger 1998: 14ff). One problem, however, continues to exist: If all systems were defined as complex systems, then the term „complexity“ and its concept would actually become meaningless. In order to avoid this problem, a qualitative discrimination becomes necessary. Two main aspects have to be taken into consideration: On the one hand, it is the level of observation and, on the other hand, it is the level of scale. Systemic complexity is a self-producing incident which only can be recognized a posteriori. In this case, the observer finds oneself on the first level of observation, she/he is part of the process and may be conscious of it, yet she/he is still unable to grasp it in its entire qualitative scope. Only after having changed the observation perspective (the second-order level), one can fully understood the connections of the processes and the formation of social, spatial and socio-spatial patterns.

At this level all the different forms of complexity can be observed – at least in principle. The level of scale refers to the various patterns of description, explication and interpretation of complex systems which vary between top-down- and bottom-up-approaches. Against this background, Richter and Rost (2002: 26) state: „Complex systems display the limitations of a reductionist perspective and have led to a change of paradigms. This anticipation, however, cannot be solved by the other extreme, that is holistic way of thinking. Chaos, for example, can be deduced from very simple and extremely reduced model systems. The key to understanding complex systems therefore lies in the selection of an adequate level of model formation and, depending on the circumstances, on the connection of different levels. This choice, however, also depends on which phenomena of a complex system are considered to be of interest" (translation A.K.).

Complex systems

Luhmann's system theory which takes sociological as well as biological theories into consideration (Maturana and Varela 1975), can be seen as an adequate basis for studying complex systems due to its terminology, theory design and transdisciplinary references. The logic of system theory allows for various systems to be analyzed separately and, at the same time, to investigate the complex interdependences of systems as synoptical hybrids resulting from coupling. What makes complex systems meaningful is already well known. Not only do natural scientists place a main focus on the research on complex structures, processes and behavioral patterns, but also economists and social scientists show an increasing interest in this approach, as can be seen, for example, in the recently published studies of Miller and Page (2007) or Epstein (2006). It is therefore not surprising that complex systems also play a significant role in geography, as has previously been stated by Ratter (2006: 112). Benenson and Torrens (2004) and Wilson (2000), just to point to two other examples, refer to space explicitly and implicitly, respectively.

An even greater challenge in geography would be to further develop current methods of modeling and simulation to render spatial complexity recognizable and comprehensible. Detecting and illustrating connections and relationships is not only challenging with regard to content, but also when it comes to the translation into simulation programs.

One has to agree with O'Sullivan (2004: 291) who states: "The entities and relationships represented are theoretical constructs concerning the subject matter, so that the computational model represents a theory about the world, rather than the world itself. The end result is a model that is itself complex, whose behavior may be almost as intractably difficult to account for as the world it represents. Connecting the model back to the world it represents is difficult for a number of reasons, principally the equifinality problem, which makes it impossible to judge the relative merits of alternative models on purely technical grounds". Nevertheless, it is crucial to visualize relationships – but to do it in another way. O'Sullivan (ibid.) offers a convincing suggestion in response to this: "This suggests a view of complexity science in which models are seen as a source of geographical narratives (stories, plots, dramas) describing how the world is, and how it might be. From this perspective it is vital that modeling is not left to modelers! Instead theories represented in models must be examined and evaluated on their own terms, [...]. It may also be useful to think of complex geographical models as extensions of thought experiments, where the necessary and contingent implications of theories can be examined".

An essential characteristic of complex systems is emergence which means that in a mostly causal process new, unprecedented and still unimaginable properties come into existence by combining various elements (Riedl 2000: 11). These unprecedented and yet non-existent characteristics must be understood relative to the respective system type. Causality seems to be no necessary property of emergence as Huber (2000: 157) states and emergence is, furthermore, not a sufficient condition for the definition of complex systems. "When we look at the behavior of a complex system as a whole, our focus shifts from the individual element in the system to the complex structure of the system. The complexity emerges as a result of the patterns of interaction between the elements" (Cilliers 1998: 5). Interaction enters the picture both as a necessary prerequisite for emergence and as a basic systemic mechanism and can be defined as the local process of the exchange of information between system elements. In other words, emergence per se cannot be regarded to be a necessary and sufficient condition for the complexity of systems (Koch 2008: 17). This necessitates further criteria which will be discussed in the following.

Discussing phenomena of emergence in geosimulation models, it is thus important to pay attention to a series of criteria, for example to the fact that emergence is an inductive bottom-up approach which has to be specified depending on the chosen system level. "We see emergent behavior in systems like ant colonies when the individual agents in the system pay attention to their immediate neighbors rather than wait for orders from above. They think locally and act locally, but their collective action produces global behavior" (Johnson 2001: 74), respectively: "Relationships in these systems are mutual: you influence your neighbors, and your neighbors influence you. All emergent systems are built out of this kind of feedback, the two-way connections that foster higher-level learning" (ibid. 120). What is true for ant colonies should not be applied to all kinds of populations in a generalized way. The relationship between emergence and interaction as described above offers further indications of fundamental characteristics of the nature of complex systems, one of them being their non-linearity, a property which has already been mentioned. Non-linearity could be described as a sort of mediator linking interaction and emergence (as inductive processes) with iteration and self-similarity (as deductive processes).

Iteration denotes the repeated transfer of a result of a function (y) at time t to a new function value (x) at time t+1. Depending on the functional context which can vary between linear, exponential, logistical, etc., different types of system impacts can be expected. The decisive question is whether the feedback loop is a self-reinforcing, positive one or a restraining and negative type. It should not be overlooked that this phenomenon of complexity, referred to as deterministic complexity, is associated with chaos. Without being able to go into detail about the fundamentals of chaos theory (for more detail, see Peitgen et al. 1994; with regard to geography, see Zibell 1995) it should be mentioned that their theoretical implications such as „iteration“, „feedback“, „path dependence“, and „sensitivity to initial conditions“, are of great importance for the understanding of complexity in systems (Manson 2001: 407ff).

The results from different epistemological approaches to complex systems – inductive by means of interaction, deductive by means of iteration – lead to the well-known idea of linking both methods to an integrative third way – the abductive approach. This is what geosimulation is currently aiming at by coupling system-dynamics-approaches with ABM-approaches. This methodological understanding offers a new and different approach to the theoretical understanding of complex systems in the con-text of social and spatial systems. The systems arising from processes of selection implemented to reduce complexity (systemic complexity), and the systems existing through processes of self-organization implemented to preserve complexity (complex systems) call attention to the processes in the respective systems. Interaction, iteration and self-organization are all concepts focusing on the system from within – they stress the operative closeness.

Complementarily, one has to explore the external systemic relationships of structural coupling which influence the operative closeness. Thus, the structural coupling is as important for the formation and development of social and spatial systems as it is the operative closeness. Formation and development, however, have to be regarded relative to what already exists, as Hard (2002: 215) argues aptly: „No matter where social geographers begin with their ‚spatial observation of social phenomena’, social systems as well as innumerable abstractions of space already exist playing an important role in social actions” (translation A.K.). Exactly this relationship is expressed by the concept of structural coupling which includes the independence of social and spatial systems; they must be tied together in order to achieve a social-spatial reality/truth.

Social and spatial systems as complex systems

Following Luhmann (1993), every social system is an autopoietic, self-referential system. The system’s elements are communication processes. Autopoiesis is the core principle on which the concept of operative closeness is based in order to reproduce the elements of a system. The justification for the thesis that social systems are autopoietic is interesting, because this idea was originally referred to living systems exclusively (Maturana 2000). Luhmann himself claims that the term „autopoiesis“ can be used in all cases in which it is possible to determine a specific operational method taking place solely in this system (Baraldi et al. 1997: 29). In so far as it can be demonstrated that spatial systems use a specific operational method, they can be defined as autopoietic systems. Communication is then the dynamic process which decides about the existence of a social system, and as long as communication follows communication, the system remains intact. The reciprocal process of interaction and iteration guarantees the stability of the system.

Therefore, one of the fundamental research challenges of geosimulation is addressed: to understand how system-stabilizing iterations and system-generating interactions mutually determine each other (Kron 2002). As far as iteration is concerned, the question arises by which means variability within the range of recursive operations is caused and how it varies, that is: how to handle the sensitivity of initial conditions. Is it a continuous fact or is it possible to interactively change its significance? As far as interaction is concerned, it may be interesting to investigate whether and in which ways social routines do relax the concept of emergent structures in such a way that the exclusive notion of the "social construction of ..." can be extended to recursive processes which include reconstructions of something. These considerations are highly relevant when exploring the purpose of social systems. In this context it can be stated that every social system reduces complexity in order to ensure that it can maintain the ability of acting as an individual in society.

This approach includes a number of further phenomena, among others the ability of collective participation or the formation and development of personal identity (Kaufmann 2005). The individual is thereby included into the social systems by operational coupling. This kind of coupling is brought forth by the reciprocal relationship of interaction and iteration. In general, the type of "operational coupling" ties together different kinds of systems, whereas systems of the same type are tied together by structural coupling (see Fig. 5). This idea can easily be applied to the coupling of different social systems with different spatial systems (Koch 2009).

In this way, the duality of systems in general and of social and spatial systems in particular, which is of utmost importance in geosimulation, can be recognized. This implies the following: social and spatial systems each are independent and at the same time interdependent. Due to the independence of space as a system it is theoretically and epistemologically possible to create models of space as self-organizational systems. The property of independence in turn needs to be associated with

Figure 5: Relationship of Structural Coupling and Emergence

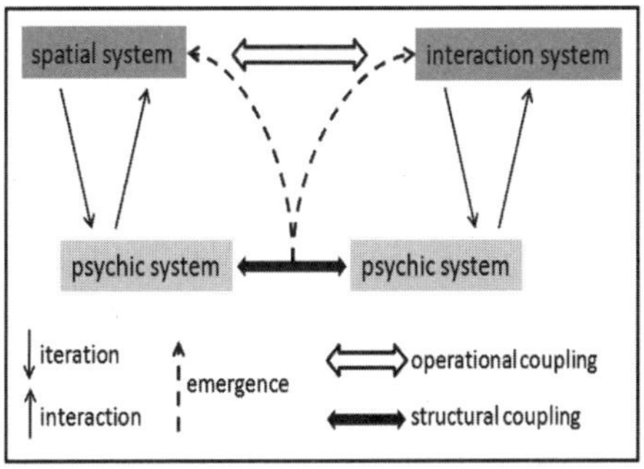

dependence complementarily, in order to manifest itself as a self-organizational and independent system. The emergence of a complex spatial system can only succeed when psychic and/or social systems need to have a reference to space; and vice versa. Consequently, the social construction of space can be interpreted as a necessary precondition of societal existence. Still, it does not suffice to answer the question as to for what purpose and in which ways space is constructed. An important step in answering these questions is to accept spatial systems as an independent dimension.

Against this background we interpret Stichweh's considerations as an appropriate alternative. Stichweh (2003: 95) criticizes the claim of a "social construction of space" as being trivial; he favors the "social control of space": "Taking into account the exteriority of space and time it appears to be fair to use the term of the social control of time and space instead of the term of the social construction of time and space. The term control means that one accepts a given fact in its independence [emphasis by A.K.]. However, subsequent operations try to subordinate this fact under the imperatives of another system. This becomes particularly evident where a strategy of controlling space is concerned, namely the invisibility of an actually given spatial order with its basal structure. This is a matter of strategies and techniques which use space but render the structuring effect of this use invisible. Different features of telecommunication media offer an example for the spatial structure and distribution of servers which serve for the Internet as infrastructure".

Löw (2008) too advocates the idea of independent spatial structures when speaking of cities as self-logical systems which constitute meaning and cognitive effects in social and mental systems. The focus should be placed on the independence of space. This independence allows for the creation of mutual relationships – this is what Löw describes when referring to „place-in-society" and „society-in-place"-approaches (ibid. 35).

Geosimulation: methodological-epistemological synthesis of a space-time-society approach

In modeling and simulating socio-spatial phenomena, one has to deal with similar problems to those arising in the simulation of solely social phenomena. Although simulation theories and techniques have undergone profound changes in the past two decades or so, several challenges remain for ongoing research efforts. In 1998, Conte, Hegselmann and Terna published a book titled "Simulating Social Phenomena". Several problems discussed at that time play a crucial role in current investigations; including questions of the adequate scaling perspective required with respect to model micro-macro structures (bottom-up vs. top-down) and their link to different social and spatial levels (Conte et al. 1998: 7). This issue relates to theory – which theoretical level is adequate to apply; is it the micro perspective as ethnography or action theory propose, or is it the macro perspective as system theory suggests? – as well as to methodology. In this case, system dynamics techniques are opposed to agent-based modeling techniques; while the former are interested in representing system behavior at an aggregated level, the latter replicates system structure as precisely as possible.

Although different objectives and/or philosophies are involved in the modeling domain, it is, nevertheless, a valid endeavor to simulate multi-level environments in order to study the abovementioned interdependencies. For this reason the micro-macro distinction supports only an initial discrimination and complex theoretical phenomena such as emergence, self-organization or (non-linear) dynamics appear to be crucial (Conte et al. 1998: 10f). The following reflections launch into this issue: as soon as effects of the mutual connectedness of the micro- and macro-levels are under consideration, theoretical and methodological implications have to be calibrated. From a theoretical perspective, it was suggested in chapter 3 to extend the theory of social systems (Luhmann 1993) to spatial systems and to integrate different system levels according to their model purpose. For instance, in analyzing socio-spatial processes of residential mobility, decisions of households play a significant role and are tied to social and spatial interactions (there is not room here to go into too much detail, so we therefore refer the reader to Benenson et al. 2002; Bruch 2006; Crooks 2008; Epstein 2006; Fossett and Senft 2004; Fossett and Waren 2005; Laurie and Jaggi 2003; Pancs and Vriend 2007; Schelling 1969, 1971). These interactions generate a specific new local pattern of social configuration (and a social pattern of local configuration) which simultaneously embeds into pre-existing relationships of social and spatial structures. The extreme poles of the micro and macro thus provide only a framework for contingent processes that occur within. The argument presented here is that a system theory which considers social and spatial systems complementarily is also capable of representing and visualizing this contingency space. The remaining question however is: what methodological-epistemological functions and purposes do spatial systems serve in the domain of geosimulation in order to achieve this theoretical claim?

From this point of view, the ultimate goal can be seen in the coupling of system dynamics and multi-agent based simulation techniques to bridge operationally bottom-up and top-down processes of interaction. Multi-level effects and results can be modeled and intuitively illustrated using such synthesized model environments. This effort is in accordance with a well-known goal: "Social theory can be significantly implemented by means of a heuristic, exploratory employment of simulation models. [...] The simulation world should be seen not only as a model of some real-life target phenomenon, but also as a potential, virtual reality which has a scientific interest of its own" (Conte et al. 1998: 14).

When focusing on the methodological impacts of modeling relationships between social and spatial systems, of which the segregation model of Salzburg is just one example, we instantaneously refer to (geo)simulation as a different and new "[...] means of knowledge production and especially representation of complex dynamics in science and technology [...]" (Küppers et al. 2006: 4), thus directly referring to the epistemological impacts of simulation. Indeed, simulation in general and geosimulation in particular can be appreciated as an alternative approach to scientific reasoning (see e.g. Rauh et al. 2008: 77f). Axelrod (1998: 24) described it as a third way of practising science: "Like deduction, it starts with a set of explicit assumptions. But unlike deduction, it does not prove theorems. Instead, a simulation generates data that can be analyzed inductively. Unlike typical induction, however, the simulated data comes from a rigorously specified set of rules rather than direct measurement of the real world. While induction can be used to find patterns in data, and deduction can be used to find consequences of assumptions, simulation can be used as an aid intuition".

Even though simulation proceeded with its epistemological standards, the underlying message is still justified in contemporary applications. In their well-known publication, Benenson and Torrens (2004: 1) defined geosimulation as being concerned "[...] with the design and construction of object-based high-resolution spatial models, using these models to explore ideas and hypotheses about how spatial systems operate [...]". By introducing and appreciating the methodological potential of geosimulation techniques they imply confirmation of a positivistic epistemology: "Here is our manifesto: We say yes to (a) positivism; [...]" (p. 252). This attitude ignores the fact that geosimulation models do not represent the reality per se; models are constructs, reducing complexity and thus making observation and perception of our environment accessible; models are also constructs accounting for scientific demands (like validity, reliability, verification), but doing so without assuming an objective reality/truth or claiming to approach this reality/truth. It is more likely that, as Ihde (2006: 84) put it: "There is no original from which to copy. Yet the end result is image-like; it is a gestalted pattern which is recognizable, although it is a constructed image".

Although the imagination of an objective reality was perhaps partially plausible in the early days of quantitative geography, it has been more or less reduced as a dominant way of thinking nowadays, without however being completely disappeared which may be seen in contemporary controversies about modeling. "Recent developments and applications of complexity theory, agent-based modeling, evolutionary and non-equilibrium-based models, Bayesian statistical inference, geocomputation, and geovisualization also suggest that the complexities of urban and social systems and people's everyday life can be taken into account in quantitative models to a certain extent [...]. Although these recent developments have moved quantitative geography away from positivist tenets [...], recent attempts to delink positivism and quantitative methodologies do not seem to be very successful; many critical geographers continue to see quantitative geography as a largely positivist and empiricist endeavor" (Kwan, M.-P. and T. Schwanen 2009: 286).

Here, we rely on Benenson and Torrens' definition because they explicitly refer to spatial systems and their operating modes. On the other hand, we disagree with positivistic epistemology and advocate a constructivist approach, because we intend to model and simulate relationships in their socio-spatial situations. As a result, neither an objective claim nor a subjective reductionism of truth or reality is made. Both levels and scales do 'collapse' and resolve in the concrete generation of a model. This is also true for the discrimination of quantitative and qualitative research: Geosimulation transforms qualitative facts into quantitative values and at the same time produces qualitative facts which have to be interpreted with regard to the concrete model at hand. In other words, the interpretation of results significantly depends on the intention of simulation: this may be an optimization strategy, sensitivity analysis, parameter comparison or computation, or any other phenomena one is interested in (see also Küppers and Lenhard 2006: 90).

The problem that has to be solved appropriately is concerned with one-sided, reductionist, mono-causal declarations. Just as space is not exclusively socially constructed, there is no exclusively external reality. In an epistemological perspective neither simple positivism nor radical constructivism can be concluded, but instead they mutually refer to one another. Most interesting perhaps, is what we can expect from this. Insisting on one-sided positions gives rise to confusing and illogical arguments: positivism would have eliminated (natural) science long ago, because the 'world' was already recognized objectively in Aristotle's times; radical constructivism has to claim itself positivistic and objective in order to remain constructivistic (else it would fail to be a kind of relativism). In addition, curious connotations have to be uncovered as Barnes (2009) or Wyly (2009) have demonstrated convincingly in following the discussion of quantitative geography. By referring to Harvey's (1973) programmatic statements which "[...] called out spatial science as neo-classical and positivist" (Wyly 2009: 313), Wyly concludes: "Unfortunately, many of these distinctions have been lost over the years, leaving a blurred collective memory of a backlash against a "positivism" that was in fact a generalization built on a very particular socially and politically constructed world. Struggles between Berry and Harvey, and among other prominent figures, personalized the conflict in vivid, memorable terms, making it all too easy to generalize from narrow, specific critiques of method or practice to more fundamental collisions of theory, epistemology, and ideology. As a result, the contextual terms of debate forged in the early 1970s were universalized into an assumption that there was something inherent and essential in the links between positivist epistemology, mathematical/statistical/quantitative methodology, and elite conservative politics" (Wyly 2009: 314). How problematic such an insistence on dichotomy can be is exemplified by Ellis' (2009) collection of racial data.

These remarks do not coincide with an attitude proclaiming one could arbitrarily be a positivist and a constructivist at the same time. Instead, the opinion held here is that extreme positions only stretch that multi-dimensional framework of cognition and knowledge in which scientific problems can be investigated. To make this framework vivid, Wyly (2009: 316) proposes a strategic positivism (and here strategic incorporates pragmatism) which "[...] recognizes the dangers of universalizing, decontextualized epistemological truth claims of the sort advocated by hardcore positivists in the mid-twentieth century". One of his proposals is hereby directly linked to our claim of adequate modeling and simulation, "[...] to translate social constructivism al dente into clear, workable research paths" (Wyly 2009: 318).

Within this framework one-sided associations become increasingly blurred. According to Barnes (2009), we can inevitably conclude that this may lead quantitative geography back to its original state, albeit at a new qualitative level. In this (historical) context, the emergence of the quantitative revolution was associated with critical politics. Barnes refers to, amongst others, Karl Marx who "[...] thought mathematics necessary to represent and understand the laws of motion of capitalism" (Barnes 2009: 294). Furthermore, location theorists like von Thünen, Weber and Christaller who rigorously relied on locational analysis, adopted mathematics "[...] for politically progressive ends, to forward social critique" (Barnes 2009: 297). This is not to deny that geosimulation is rooted in quantitative geography, but it does not however, imply either a radical positivistic methodological embeddedness or a missing potential critical with respect to simulated results. Critical power refers to people and theories, not only to methods.

Nevertheless, questions about the appropriate modus operandi with which to mutually relate quantitative and qualitative methods in a post-positivistic era remain. In other words: can we succeed in translating between the digital and analogue worlds of language? By considering post-structuralism and GIScience comparably, Leszczynski (2009) observes quantitative limits to qualitative engagements. She states that a successful translation between the two spheres is simply impossible due to their fundamental differences. "What I term here the philosophical divide is characterized in terms of two very different discourses: the digital universe to which GIS belongs and the linguistic realm of (presently) poststructuralist-dominated critical theory" (Leszczynski 2009: 351). This divide does not, however, exclude hybrid research strategies that tie quantitative and qualitative methods with reasoning. Moreover, critical "[...] GIS practices certainly employ and constitute qualitative methods, but these are distinctly quantitative qualitative engagements that are ultimately expressed within the universe of computing" (ibid. 362).

The divide is first and foremost justified by a logical opposition of ontology and epistemology. While GIScience basically conceptualizes knowledge ontologically, this is also the status of epistemology in the realm of qualitatively and poststructuralist grounded linguistic approaches. "In other words, GIS is epistemologically insufficient because it objectives knowledge" (Ahlqvist 2000 in Leszczynski 2009: 356). On the other hand, human beings are free to infer meaning from context. At this very general level one can agree with her in as much as GISystems – as well as geosimulation models – are not capable of self-reflection and of asking fundamental philosophical questions as human beings can do. But if one is stepping deeper into the subject matter then some doubts may arise.

Referring to the distinction between 'infological' and 'datalogical' made by Feuchtwanger and Poiker (1987), Leszczynski (2009: 357) posits that: "Infological representations constitute semantic models of how the real world is conceptualized and represented. Here, ontological commitments are made via the enumeration of the phenomena that "may exist and how they fit together" [...]. In the datalogical world, the "same things" are defined, "but in computer terms" [...]. Datalogical representation therefore concerns how conceptual models are organized in the computer. This includes formal schema and the computer architectures used to register – or physically store – spatial representations". GIS is in this sense a closed system of conceptualized and formalized language, a formal realm of discourse. If we consider consciousness as a computer-like machine then the relationship between infological (semantics) and datalogical (syntax) worlds is true even for human language (as it is for mathematics). Obviously, GIS and geosimulation are far from being as flexible in coupling syntax and semantics as human brains are – but these are differences of scale or size and not of category.

With respect to translations not within but between these systems (cognition and computer), we agree with Leszczynski's conclusion about asymmetries between the two worlds: "All representation begins in the infological world as concepts. It is only subsequently defined via translation to the datalogical universe via mapping elements of these concepts to digital sets" (ibid. 358). Translations of that kind induce a semantic loss due to the intrinsic properties of the digital world. At least three questions are attached to this context. Firstly, the re-translation mechanism is missing. Spatial representations of the cognitive-semantic world are probably more complex and diverse than those of the GIS world. On the other hand, it is exactly this kind of representations – be it maps, aerial photos, satellite images or navigational systems – that significantly influence how we spatially recognize our environment. This is a mutual relationship. Secondly, contemporary developments in GIScience (one may think of Geoweb 2.0, geo-social networks or virtual globes) have led to a tremendous extension of the datalogical world – and in light of, for example, spatial data mining (e.g. Behnisch 2008; Ultsch 2003) or pervasive and grid computing (e.g. Lanig et al. 2008; Padberg and Greve 2009), this process seems to be open-ended. Thirdly and finally, specific progress in geosimulation technologies also accounts for a different perspective. This is well illustrated by attempts to generate emergent macro-structures out of local rules, as urban segregation research, game-of-life approaches or Wolfram's (2002) 1D- and 2D-universes show. A related branch is the integration of processes of self-organization with genetic algorithm or evolutionary programming techniques (e.g. Hesse 2005; Michalewicz 1996; Miller and Page 2007). From this perspective, I would argue conclusively that it no longer seems to plausible to support Gould's (1999) statement in its totality that "[…] attempts to abstract the social world into the closed system of mathematics not only yield absurdity but are counterintuitive as they negate the very purpose of mathematization, which is predictive description (of some system or outcome). Prediction can only be achieved with advance knowledge of possible outcomes; the human and social dimensions, however, introduce infinite degrees of complexity and chaos that cannot be accommodated or captured even by nonlinear stochastic systems such that knowledge in advance will never be possible[…]" (Gould 1999, in: Leszczynski 2009: 354).

Concluding remarks

At the preliminary end of our expedition to the theoretical, methodological and epistemological objectives of the geosimulation universe, we are able to recognize some of the relationships between spatiality, temporality and sociality which should be taken into account when investigating social life with modeling and simulation techniques. The central message is that space – including all those notions that are referred to as, for example, spatiality, place, location or network –influences social structures and individual actions. The implications are manifold and diverse. First and very generally, it is necessary to coherently and precisely incorporate theoretical requirements associated with the model's purpose. This relationship with theory appears to be crucial to the reliability of the methodological instruments applied in simulation. This is true even if a pragmatic access to reality and truth is chosen. In the light of a constructivist epistemology, such a diverse approach to reality is nevertheless possible, excluding the extreme positions of both theory-free and theory-determining modeling beliefs. In this sense we can confirm that simulation is keen on stimulation, as Resnick (1997) emphasizes, but also includes imitation and prediction as well: "I'm not interested in developing accurate imitations of these real-world systems, or even in making accurate predictions about them. The real world serves only as an inspiration, a departure point for thinking about decentralized systems. When I write a StarLogo program with artificial ants, for instance, I am more interested in investigating antlike behaviors than the behavior of real ants. Even more, I am interested in how people think about antlike behaviors. In short, I am more interested in stimulation than in simulation" (Resnick 1997, 49).

The justification for considering space explicitly has been unfolded in this article. It is not only geography in general and geosimulation in particular that are interested in this argument – Nagel and Koll-Schretzmayr (2007), Rindsfüser and Klügel (2007) and Semboloni (2007) for example, illustrate convincingly from a micro-spatial perspective that it is in fact crucial to include space in models of personal behavior. Game theory can also be mined for useful advice for thinking in and with spaces. Majeski et al. (1998: 166) for instance concluded from their spatially-iterated prisoners dilemma game simulations, comparing experiments with and without consideration of space, that "[m]ovement is an important factor in the development of a stable cooperative world because it allows actors to change location and find more cooperative neighbors and because it reduces the number of captive prey for exploitative strategies, making their own survival more precarious".

System theory as it is presented here, as an interdependent interweaving of social and spatial systems, is – with respect to efforts of balancing the dynamics between social and spatial structures – quite similar to other proposals such as the space syntax approach introduced by Hillier and Hanson (1984), because mutuality is here also put to the fore. Portugali (2006: 18), in his complex artificial environment approach, also refers to it: "Hillier and Hanson further suggest three sets of laws, according to which artificial environments such as cities evolve: laws regarding the relation between spatial objects, "laws from society to space" and "laws from space to society" […]". If this idea is accepted then we must deal with three different types of interaction, namely (a) among social agents, (b) among spatial agents and (c) between social and spatial agents. Consequently, complexity (theory) actually becomes complex, because the unavoidable process of "simplifying complexity" must be coupled with the unavoidable process of "complexifying simplicity".This is true not only for the world of geosimulation but also seems to be a common issue in quantitative geography, as Zolnik (2009: 339) points out: "The emergence of local spatial analysis in human geography to account for spatial non-stationarity mirrors the evolution of quantitative methods in the social sciences from parsimony – as simple as possible models – to realistic complexity – as complex as necessary models".

According to system theory and its implications for (human) geography, the still existing missing link is that which Portugali (2006: 16) describes as 'circular causality': a unified embeddedness of micro/local and macro/global structures and processes. Research into self-organization and emergence unfolds its full meaning only if their implications are complementarily involved, i.e. in the implementation of emerged, top-down effecting structures. Additionally and much more complex, is the differential mutual coupling of variously scaled spaces and societies. The question, nevertheless, remains: what is 'top' and what is 'bottom' in all these relationships?

One final comment is dedicated to the different kinds of knowledge developed in the methodological and geographical disciplines, and the necessity to incorporate both universes adequately, even if the latter is less formalized. Pumain (2006: 73) puts it in a nutshell: "I want to develop [...] the complementary idea that, while the so-called (self-called?) "sciences of complexity" are evolving towards an attempt at unifying the analysis and, even if possible, "the" theory, of complex systems, social scientists should be keen on maintaining their previous knowledge as a most valuable input in the models that are now developed. Actually, that knowledge about social systems, although less formalised and incompletely integrated, already incorporates the principles that are today the distinguishing markof complexity theory, and then should be recognized and integrated as such: heterogeneity of elements and their properties, diversity of interactions [...], path dependency of the evolution, unpredictability of the future [...]". A geosimulation technique that pays attention to this idea would represent significant progress on the way to transdisciplinarity and relational thinking.

References

Axelrod, R. (1998): Advancing the Art of Simulation in the Social Sciences. In: Conte, R., Hegselmann, R. & P. Terna (eds.) (1998): Simulating Social Phenomena. Springer Verlag, Berlin, Heidelberg, New York, pp. 21-40.

Baraldi, C., Corsi, G. u. E. Esposito (1997): GLU – Glossar zu Niklas Luhmanns Theorie sozialer Systeme. Frankfurt a. M.

Barnes, T. (2009): "Not Only ... But Also": Quantitative and Critical Geography. In: The Professional Geographer, 61(3), pp. 292-300.

Bartelme, N. (2005): Geoinformatik. Modelle, Strukturen, Funktionen. 4. Auflage, Springer Verlag, Berlin.

Batty, M. (2005): Cities and Complexity. Understanding Cities with Cellular Automata, Agent-Based Models, and Fractals. Cambridge, Massachusetts.

Behnisch, M. (2008): Urban Data Mining. Operationalisierung der Strukturerkennung und Strukturbildung von Ähnlichkeitsmustern über die gebaute Umwelt. Universitätsverlag Karlsruhe.

Benenson, I. & P. M. Torrens (2004): Geosimulation: Automata-based modeling of urban phenomena. John Wiley & Sons, Chichester UK.

Benenson, I., Omer, I., & E. Hatna (2002): Entity-based modeling of urban residential dynamics: the case of Yaffo, Tel Aviv. In: Environment and Planning B, vol. 29, pp. 491-512.

Bossel, H. (2004): Systeme, Dynamik, Simulation. Modellbildung, Analyse und Simulation komplexer Systeme. Norderstedt.

Bruch, E. (2006): Residential Mobility, Income Inequality, and Race/Ethnic Segregation in Los Angeles. In: Population Association of America, 2006 Annual Meeting, available at: http://paa2006.princeton.edu/download.aspx?submissionId=60143 (23.10.2009)

Cassirer, E. (1931): Mythischer, ästhetischer und theoretischer Raum. In: Dünne, J. & S. Günzel (eds.): Raumtheorie. Suhrkamp Verlag, Frankfurt a. M., pp. 485-500.

Cilliers, P. (1998): Complexity & Postmodernism. Understanding complex systems. London und New York.

Conte, R., Hegselmann, R. & P. Terna (eds.) (1998): Simulating Social Phenomena. Springer Verlag, Berlin, Heidelberg, New York.

Crooks, A. (2008): Constructing and Implementing an Agent-Based Model of Residential Segregation through Vector GIS. In: UCL Working Papers Series, Paper 133.

De Lange, N. (2005): Geoinformatik in Theorie und Praxis. 2. Auflage, Springer Verlag, Berlin.

Doetsch, H. (2006): Einleitung. In: Dünne, J. & S. Günzel (eds.): Raumtheorie. Suhrkamp Verlag, Frankfurt a. M., pp. 195-211.

Dünne, J. (2006): Einleitung. In: Dünne, J. & S. Günzel (eds.): Raumtheorie. Suhrkamp Verlag, Frankfurt a. M., pp. 289-303.

Ellis, M. (2009): Vital Statistics. In: The Professional Geographer, 61(3), pp. 301-309.

Epstein, J. (2006): Generative Social Science. Studies in Agent-Based Computational Modeling. Princeton and Oxford, 2006.

Feuchtwanger, M. & T. K. Poiker (1987): Infological and datalogical spatial models. In: Lecture notes in digital mapping and land information, Department of Surveying Engineering, University of Calgary, Canada, pp. 146-167.

Fischer, F. (2009): Volunteered Geographic Information – Baustein zukünftiger Geoinformationsinfrastrukturen? In: Wiener Schriften zur Geographie und Kartographie, Band 19, Wien, pp. 148-153.

Fischer-Kowalski, M. u. K. Erb (2006): Epistemologische und konzeptuelle Grundlagen der Sozialen Ökologie. In: Mitteilungen der Österreichischen Geographischen Gesellschaft, Band 148, pp. 33-56.

Fliedner, D. (2001a): Der anthropogene Raum, Gegenstand einer handlungs- oder prozeßorientierten Sozialgeographie? In: http://www.geographische-revue.de/archiv/fliedner.pdf (01.02.2008), pp. 1-26.

Fliedner, D. (2001b): Six Levels of Complexity; A Typology of Processes and Systems. In: Journal of Artificial Societies and Social Simulation, vol. 4, no. 1 http://www.soc.surrey.ac.uk/JASSS/4/1/4.html, (01.0.2008) pp. 1-17.

Fliedner, D. (1999): Komplexität und Emergenz in Gesellschaft und Natur. Frankfurt a. M.

Flusser, V. (1991): Räume. In: Dünne, J. & S. Günzel (eds.): Raumtheorie. Suhrkamp Verlag, Frankfurt a. M., pp. 274-285.

Fossett, M. & R. Senft (2004): SimSeg and Generative Models: A Typology of Model-Generated Segregation Patterns. In: Proceedings of the 2004 Conference on Social Dynamics: Interaction, Reflexivity, Emergence, pp. 39-78.

Fossett, M. & W. Waren (2005): Overlooked Implications of Ethnic Preferences for Residential Segregation in Agent Based Models. In: Urban Studies (42) 11, pp. 1893-1917.

Foucault, M. (1984): Von anderen Räumen. In: Dünne, J. & S. Günzel (eds.): Raumtheorie. Suhrkamp Verlag, Frankfurt a. M., pp. 317-329.

Goodchild, M. (1992): Geographical Information Science. In: International Journal of Geographical In-formation Systems, vol. 6, no. 1, pp. 31-45.

Hard, G. (2002): Auf der Suche nach dem verlorenen Raum. In: Landschaft und Raum. Aufsätze zur Theorie der Geographie, Band 1 (=Osnabrücker Studien zur Geographie, Band 22), Osnabrück, pp. 211-234.

Hesse, R. (2005): Ein Hybrid-Verfahren zur Bearbeitung Kombinatorischer Optimierungsprobleme. Die Integration von Multiagentensystemen und Genetischen Algorithmen in einem Framework zur Behandlung von Optimierungsproblemen aus der Praxis. Dissertation TU München.

Huber, L. (2000): Wie das Neue in die Gehirne kommt. Emergenz und Chaos in neuronalen Prozessen. In: Huber, L. (Hrsg.): Wie das Neue in die Welt kommt. Wien, pp. 157-175.

Ihde, D. (2006): Models, Models, Everywhere. In: Lenhard, J., Küppers, G. & T. Shinn (eds.): Simulation. Pragmatic Construction of Reality. Springer Verlag, Dordrecht, pp. 79-86.

Johnson, S. (2001): Emergence. The connected lives of ants, brains, cities, and software. New York, London, Toronto, Sydney.

Kaufmann, J.-C. (2005): Die Erfindung des Ich. Eine Theorie der Identität. Konstanz.

Klüter, H. (1999): Raum und Organisation. In: Meusburger, P. (Hrsg.): Handlungszentrierte Sozialgeographie. Benno Werlens Entwurf in kritischer Diskussion, pp 187-212.

Klüter, H. (1986): Raum als Element sozialer Kommunikation. (=Giessener Geographische Schriften), Heft 60. Giessen.

Koch, A. (2010): Ein multivariates Segregationsmodell mit GIS und ABM. In: Strobl/ Blaschke/Griesebner (eds.), Heidelberg 2010, (paper accepted).

Koch, A. (2009a): Agents in Space meet Spatial Agents. Modeling segregation as a complex spatially explicit phenomenon. In: European Journal of GIS and Spatial Analysis, vol. 19, no. 4/2009, pp. 471-491.

Koch, A. (2009b): Zur Komplementarität von struktureller Kopplung und Emergenz. Ein Vorschlag zur systemtheoretischen Einbettung des Räumlichen für eine integrative Geographie. In: Koch, A. (eds): Mensch-Umwelt-Interaktion. Überlegungen zum theoretischen Verständnis und zur methodischen Erfassung eines grundlegenden und vielschichtigen Zusammenhangs. Salzburger Geographische Arbeiten, Band 45, Selbstverlag Salzburg, pp. 47-56.

Koch, A. (2008): Modellierung und Simulation komplexer geographischer Systeme. Eine systemtheoretische Betrachtung zur Komplexität sozialer und räumlicher Systeme. In: Mandl, P. & A. Koch (eds.): Modellierung und Simulation komplexer geographischer Systeme. Salzburger Geographische Arbeiten, Band 43, Selbstverlag Salzburg, pp. 9-31.

Koch, A. (2004): Dynamische Kommunikationsräume. Ein systemtheoretischer Raumentwurf. LIT Verlag, Münster.

Krieger, D. J. (1998): Einführung in die allgemeine Systemtheorie. 2. Aufl., München.

Kron, Th. (Hrsg.) (2002): Luhmann modelliert. Sozionische Ansätze zur Simulation von Kommunikationssystemen, Opladen.

Küppers, G., Lenhard, J. & T. Shinn (2006): Computer Simulation: Introductory Essay. In: Lenhard, J., Küppers, G. & T. Shinn (eds.): Simulation. Pragmatic Construction of Reality. Springer Verlag, Dordrecht, pp. 3-22.

Küppers, G. & J. Lenhard (2006): From hierarchical to network-like integration: a revolution of modeling style in computer-simulation. In: Lenhard, J., Küppers, G. & T. Shinn (eds.): Simulation. Pragmatic Construction of Reality. Springer Verlag, Dordrecht, pp. 89-106.

Kwan, M.-P. & T. Schwanen (2009): Quantitative Revolution 2: The Critical (Re) Turn. In: The Professional Geographer, 61(3), pp. 283-291.

Lanig, S., Schilling, A., Stollberg, B. & A. Zipf (2008): Towards Standards-Based Processing of Digital Elevation Models for Grid Computing through Web Processing Services (WPS). In: Gervasi, O. et al. (eds.): ICCSA 2008, Part II, LNCS 5073, pp. 191-203.

Laurie, A. J., N. K. Jaggi (2003): Role of 'Vision' in Neighborhood Racial Segregation: A Variant of the Schelling Segregation Model. In: Urban Studies, vol. 40, no. 13, pp. 2687-2704.

Lefebvre, H. (1991): The Production of Space (translated by Donald Nicholson-Smith). Blackwell Publishing, Malden (USA), Oxford (UK), Carlton (Australia).

Lefebvre, H. (1974): Die Produktion des Raums. In: Dünne, J. & S. Günzel (eds.): Raumtheorie. Suhrkamp Verlag, Frankfurt a. M., pp. 330-342.

Leszczynski, A. (2009): Quantitative Limits to Qualitative Engagements: GIS, Its Critics, and the Philosophical Divide. In: The Professional Geographer, 61(3), pp. 350-365.

Löw, M. (2008): Soziologie der Städte. Suhrkap Verlag, Frankfurt a.M.

Longley, P. A. (2010): Geographic Information Systems and Science. John Wiley & Sons, Chichester UK.

Luhmann, N. (1993): Soziale Systeme. 3. Auflage, Suhrkamp Verlag, Frankfurt a.M.

Majeski, S., Linden, G., Linden, C. & A. Spitzer (1998): A Spatial Iterated Prisoners Dilemma Game Simulation with Movement. In: Conte, R., Hegselmann, R. & P. Terna (eds.) (1998): Simulating Social Phenomena. Springer Verlag, Berlin, Heidelberg, New York, pp. 161-167.

Mandl, P. (2003): Multi-Agenten-Simulation und Raum – Spielwiese oder tragfähiger Modellierungsansatz in der Geographie? In: Koch, A. & P. Mandl (eds.): Multi-Agenten-Systeme in der Geographie. Klagenfurter Geographische Schriften, Heft 23, Selbstverlag Klagenfurt, pp. 5-34.

Manson, S. (2001): Simplifying complexity: a review of complexity theory. In: Geoforum, vol. 32, pp. 405-414.

Maturana, H. R. (2000): Biologie der Realität. Frankfurt am Main.

Maturana, H. R. u. F. G. Varela (1975): Autopoietic Systems. A characterization of the living organization. Urbana-champaign, Illinois.

Michalewicz, Z. (1996): Genetic Algorithms + Data Structures = Evolutionary Programs. Third, revised and extended edition. Springer Verlag, Berlin, Heidelberg, New York.

Miller, J. H. & S. E. Page (2007): Complex Adaptive Systems. An Introduction to Computational Models of Social Life. Princeton University Press, Princeton and Oxford.

Nagel, K. u. M. Koll-Schretzenmayr (2007): Multiagenten-Simulation in der Raumplanung. In: disP 170, Heft 3, pp. 3-8.

O'Sullivan, D. (2004): Complexity science and human geography. In: Transactions of the Institute of British Geographers, vol. 29, pp. 282-295.

Padberg, A. & K. Greve (2009): Gridification of OGC Web Services: Challenges and Potentials. In: GIS.Science 14, no. 3, pp. 77-81.

Pancs, R., N. J. Vriend (2007): Schelling's spatial proximity model of segregation revisited. In: Journal of Public Economics 91, pp. 1-24.

Peitgen, J., Jürgens, H. u. D. Saupe (1994): Chaos. Bausteine der Ordnung. Berlin, Heidelberg.

Ratter, B. (2006): Komplexitätstheorie und Geographie – en Beitrag zur Begründung einer anderen Sicht auf Systeme. In: Mitteilungen der Österreichischen Geographischen Gesellschaft, Band 148, pp. 109-124.

Rauh, J., Schenk, T. A. & D. Schrödl (2008): Agentenbasierte Simulation von Konsumentenhandelen. In: Mandl, P. & A. Koch (eds.): Modellierung und Simulation komplexer geographischer Systeme. Selbstverlag Salzburg, pp. 77-104.

Reitsma, F. (2003): A response to simplifying complexity. In: Geoforum, vol. 34, pp. 13-16.

Resnick, M. (1997): Turtles, Termites, and Traffic Jams. Explorations in massively parallel microworlds. Cambridge, Massachusetts.

Rheinberger, H.-J. (2001): Experimentalsysteme und epistemische Dinge. Eine Geschichte der Proteinsynthese im Reagenzglas. Göttingen.

Richter, K. u. J.-M. Rost (2002): Komplexe Systeme. Frankfurt am Main.

Riedl, R. (2000): Wie wohl das Neue in die Welt kommt? In: Huber, L. (Hrsg.): Wie das Neue in die Welt kommt. Wien, pp. 9-23.

Rindsfüser, G. u. F. Klügl (2007): Agent-Based Pedestrian Simulation. A Case Study of the Bern Railway Station. In: disP 170, Heft 3, pp. 9-18.

Schelling, T. C. (1971): Dynamic models of segregation. In: Journal of Mathematical Sociology 1 (2), pp. 143-186.

Schelling, T. C. (1969): Models of segregation. In: American Economic Review, Papers and Proceedings 59, 1969, pp. 488-493.

Seel, M. (2008): Theorien. S. Fischer Verlag, Frankfurt a.M.

Semboloni, F. (2007): The Management of Urban Complexity through a Multi-Agent Participatory Simulation. In: disP 170, Heft 3, pp. 57-70.

Stichweh, R. (2003): Raum und moderne Gesellschaft. Aspekte der sozialen Kontrolle des Raums. In: Krämer-Badoni, Th. u. K. Kuhm (Hrsg.): Die Gesellschaft und ihr Raum. Raum als Gegenstand der Soziologie. Band 21, Opladen, pp. 93-102.

Stichweh, R. (1998): Raum, Region und Stadt in der Systemtheorie. In: Soziale Systeme, Zeitschrift für soziologische Theorie, Jg. 4, Heft 2, pp. 341-358.

Strobl, J. (2009): Neogeography – globale, verteilte, kollaborative raumbezogene Information als neue Herausforderung für die geographische Forschung. In: Wiener Schriften zur Geographie und Kartographie, Band 19, Wien, pp. 107-111.

Ultsch, A. (2003): Maps for the Visualization of high-dimensional Data Spaces. In: Proceedings Workshop on Self-Organizing Maps (WSOM 2003), Kyushu, Japan, pp. 225-230.

Wilensky, U. (1999): NetLogo. http://ccl.northwestern.edu/netlogo/. Center for Connected Learning and Computer-Based Modeling, Northwestern University, Evanston, IL.

Wilson, A. G. (2000): Complex Spatial Systems. The Modeling Foundations of Urban and Regional Analysis. Harlow.

Wirth, E. (1979): Theoretische Geographie. Grundzüge einer Theoretischen Kulturgeographie. Stuttgart.

Wolfram, S. (2002): A New Kind of Science. Wolfram Media, Champaign.

Wyly, E. (2009): Strategic Positivism. In: The Professional Geographer, 61(3), pp. 310-322.

Zibell, B. (1995): Chaos als Ordnungsprinzip im Städtebau. Ansätze zu einem neuen Planungsverständnis. (=ORL-Bericht 99/1995), Zürich.

Zolnik, E. J. (2009): Context in Human Geography: A Multilevel Approach to Study Human-Environment Interactions. In: The Professional Geographer, 61(3), pp. 336-349.

An agent-based simulation to show the effects of the size of a city on the socio-spatial organisation of its population

Reinhard König

Introduction

The interrelation between the socio-spatial organisation of a population and the built structure of a city cannot be empirically verified due to lack of micro data. Simulation models offer an alternative study approach. However the results of such models strongly depend on the chosen initial assumptions. Simulation models used to investigate the process of residential segregation explore the effects of the population's tolerance towards different kinds of residents or of rents or land prices (Benenson 1998; Portugali 2000). One disadvantage of most segregation models is that the representation of the city's built structure is oversimplified and various built realities are abstracted. The model presented in this paper is a combination of a graph-based spatial representation and an agent-based simulation. It is designated as Circle City and enables us to study the effects of the size of a settlement on the socio-spatial organisation of its population.

Circle City

The simulation program was written in Delphi (Object Pascal) and compiled as a Windows executable program with a graphical interface (Figure 1). The mo-

Figure 1: Screenshot of Circle City.

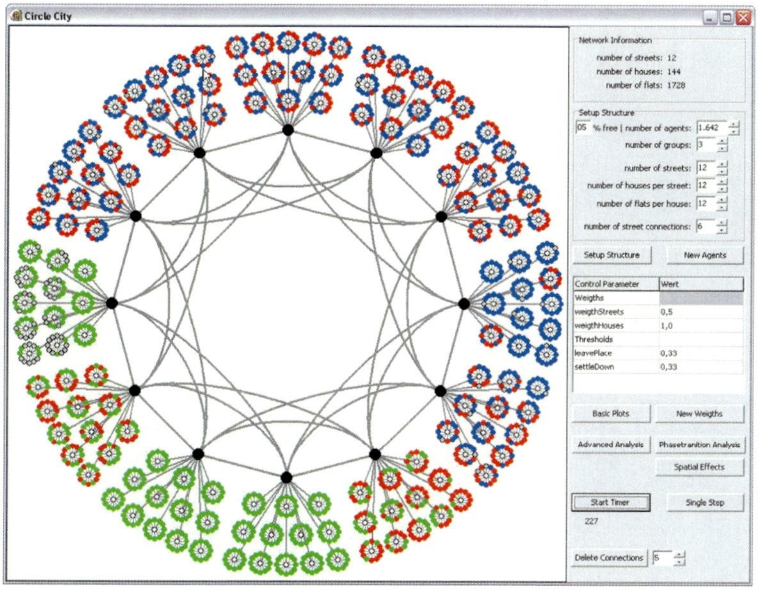

del description follows the ODD protocol for describing individual- and agent-based models (Grimm and Railsback 2005; Grimm, Berger et al. 2006) and consists of seven elements. The first three elements provide an overview; the fourth element explains general concepts underlying the model's design, while the remaining three elements provide details.

■ Purpose

In the present study we begin with the thesis that the size of a city has an essential effect on the residential segregation of the population.Various simulation experiments presented in the subsequent chapters support this assumption.

State variables and scales

With the presented model, a true-to-life representation of the built urban structure is represented by means of a graph-based method. The nodes of the graph represent the street segments, the houses and the dwellings. An edge is drawn between two street-nodes if they are connected to one another with a crossroads, a house-node is connected to a street-node if the house has its address in that street, and a dwelling-node is connected to a house-node if the dwelling is located in the house (Figure 2). With this representation we can construct a developing structure of the streets via buildings to dwellings (Figure 3). The underlying idea is that neighbours who live in the same house and who thus use the same stairwell or corridor, are more important for each other than neighbours living in an adjacent house, meeting each other in a minor road that is more a public place than a stairwell or corridor. The neighbours in the next road are again less important and, if there is a main road between two residents, the relation of these people is very slight. These different spatial levels are marked in Figure 3 with the letters a to f, where each node represents a particular hierarchical level from the individual dwelling to the motorway.

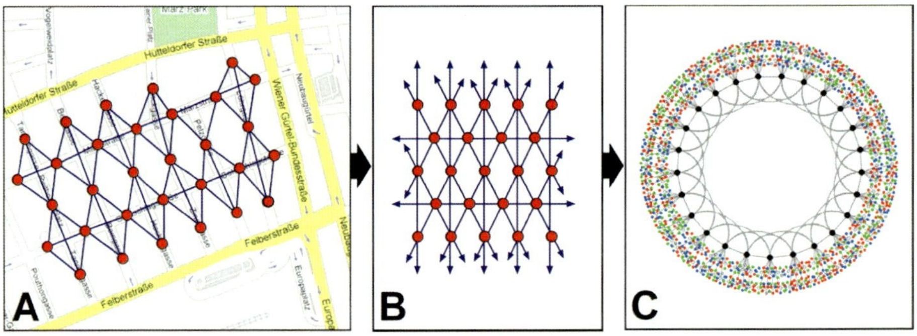

Figure 2: Design of Circle City. Translation of an urban development structure to an abstract model. A) Representation of streets and crossroads by nodes and edges. B) Graph-based representation of a gridded street network with regular intervals. C) Graphical illustration of a hierarchical network. Streets are drawn as solid black nodes, houses as white nodes, and dwellings as grey nodes with black outlines if unoccupied or coloured nodes if occupied by an agent.

This graph-based representation of a city is designated as Circle City because of its circular rimless structure, which enables the abstract design of every possible spatial configuration

Figure 3: Hierarchy of the urban development structure. The abstract hierarchical network represents the development structure from a dwelling to a motorway.

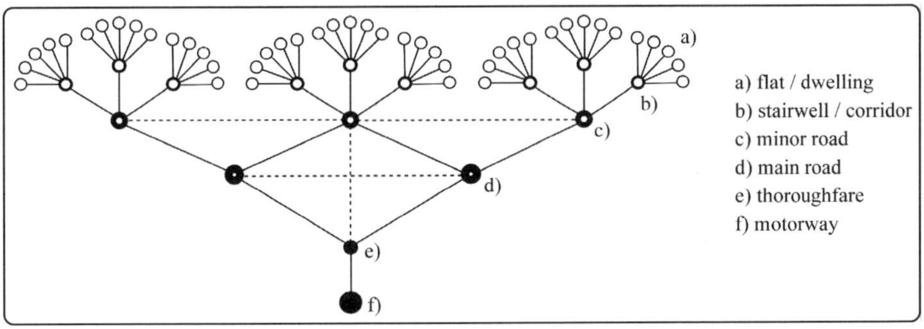

Figure 4: Weightings. The graphs a, b, c and d show different neighbourhood relations of two cells H1 and H2 with the corresponding weights wi per hierarchical level.

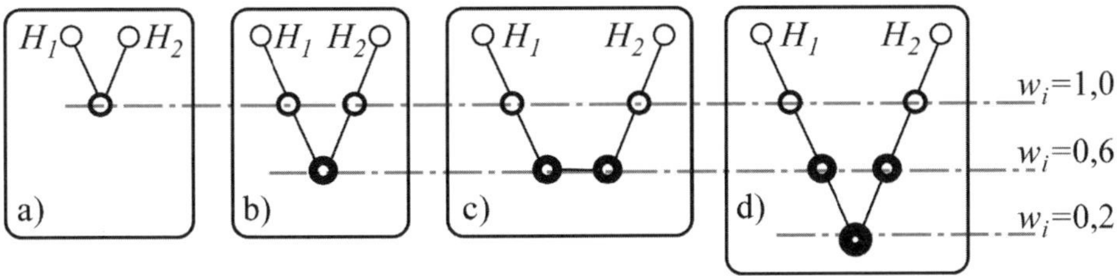

(Figure 1). The hierarchy of the urban development structure represented by this model enables a detailed examination of the effects of various arrangements of dwellings, houses, streets, and their interconnections. Differently built structures can be captured by means of various weightings between the areas of development (Figure 4). These weightings represent the different intensities of how inhabitants perceive their neighbourhood via various areas of development.

On basis of the spatial development structure that is represented by a graph, an agent-based segregation model is introduced (Figure 2, C). In the simulation of residential segregation the households are represented by agents.

Process overview and scheduling

The behaviour of agents is oriented on Thomas Schelling's fundamental model (1969). According to this model, each household tries to get a dwelling with a local neighbourhood that satisfies its requirements.

The model process is illustrated by the flow diagram in Figure 5. In the first step, the shortest paths from all nodes to all other nodes in the graph are calculated and saved. For calculating, the weightings of the nodes on the shortest path serve f the influence of one node on another. In the next step, the neighbourhood condition of all agents is calculated by looking at their group membership. Following this calculation, the agents are considered in a random order. If an agent is not satisfied with its current neighbourhood condition, it is added to an 'unhappy-list', otherwise the agent stays happy at its current location. After considering all agents, the ones of the unhappy list are called in random order. Each one tries to find a new location that fulfils its requirements. If an agent finds a satisfying location it moves to it, otherwise it stays unhappy at its old location. After considering all agents, the discrete model time is increased and the process starts again. The correspondence of one time-interval in the model to real time is not defined, since we do not try to reproduce a real residential structure, but are only interested in the model dynamics if the graph-size is manipulated.

Design concepts

Emergence: The arrangement of the different population groups on the macro level emerges from the migration dynamics which result from the individual tolerance thresholds of the households. This phenomenon is well known from Schelling's tipping point theory, which is based on a gridded space environment (Schelling 1972). In the presented Circle City, one can observe similar segregation patterns, but with a hierarchical structure that is based on streets and houses (cp. Figure 1). A mechanism for adaptation is not modelled. The agents can only adapt indirectly to the environment by migrating to another location, but they cannot leave the city, die or reproduce.

Figure 5: Flow diagram of model processes.

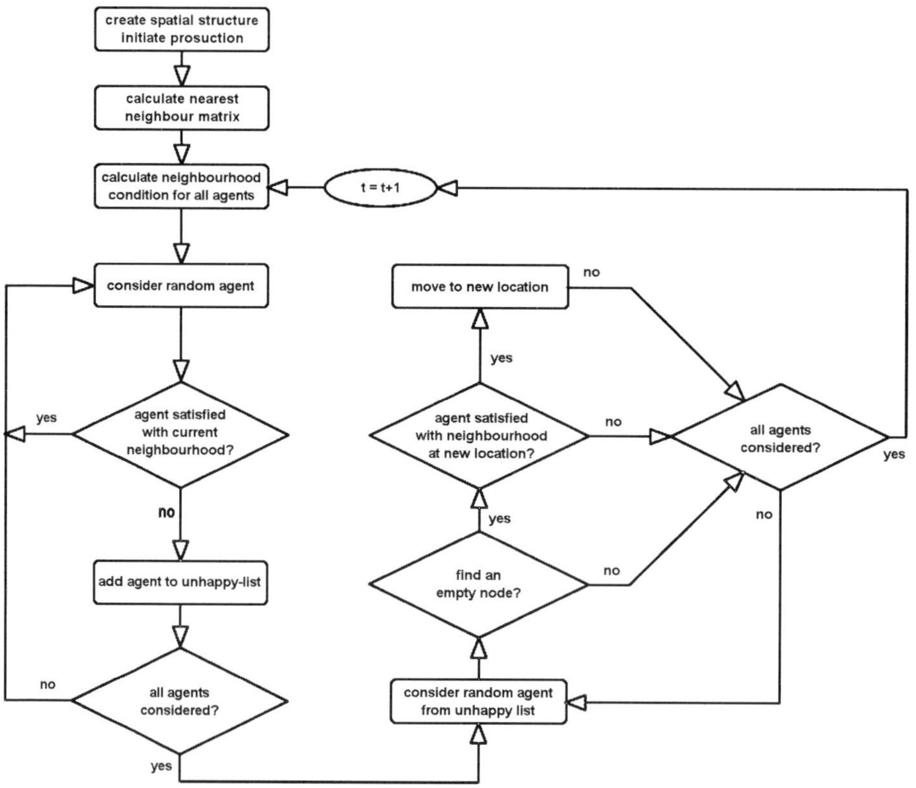

Sensing: Each agent counts the members of its group and other groups. The neighbourhood that is considered essentially comprises the whole city. This tally is the basis for calculating the neighbourhood condition.
Interaction: The agents interact via the landscape of the graph to sense the neighbourhood condition. If an agent is not satisfied with its neighbourhood, it can migrate to another location.
Stochasticity: The agents are considered in random order, both to test their satisfaction with their current location and to look for a new location. For the tolerance threshold, a discreet choice is used in contrast to the probability choice that is used, for example, by Benenson (1998). To avoid artefacts in the simulation, the order of the agents is chosen randomly at each time increment (Epstein and Axtell 1996).
Collectives: The households are each assigned to a specific group. An agent's group membership is recorded by its state parameter and cannot be changed during a simulation. All agents of all groups have the same tolerance threshold. The nodes of the graph are organised in three categories: street-nodes, house-nodes, and dwelling-nodes. All streets contain the same number of houses and all houses contain the same number of dwellings (cp. Figure 1). All nodes from a particular category have the same weights.
Observation: To obtain the degree of residential segregation of the populati-

Table 1: Variables of Circle City.

Parameter	Initial Value	Brief description
Spatial structure		
w_i	-	Weighting of a node that indicates how communicative a space is; Values range between 0 (not communicative) and 1 (very communicative)
w_h	1.0	Weighting of the house-nodes
w_s	0.5	Weighting of the street-nodes
ConnList		List of nodes that are connected to the one being considered
H		Index of a node. $H=\{1, 2, \ldots, i\}$
A		Index of an agent. $A = \{A_1, A_2, \ldots, A_{NA}\}$
Households		
Ψ_{leave}	0.33	Threshold of an agent to leave its current location; Values range between 0 (very tolerant households) to 1 (very intolerant households)
Ψ_{settle}	0.33	Threshold of an agent to settle down on a new location; Values range between 0 (very tolerant households) to 1 (very intolerant households)
L^A	$L^A \subset H$	Location (node) of an agent; Contains the identification number of a node
S^A	-	State of an agent. Indicates its group membership. $S^H=\{S_1, S_2, \ldots, S_{NG}\}$
V^A	-	Ratio of agents with the same group membership to agents with other group membership in the neighbourhood of one agent (node), taking into consideration the weightings of the nodes; The value is in the interval [0, 1]
$U(A)$	-	Neighbourhood of an agent
Higher-level entities		
N_A	95%	Number of agents in a simulation; Upon initialisation of a simulation, 95% of the nodes are occupied by agents
N_G	3	Number of groups into which the agents are divided
N_s	12	Number of streets in a simulation
N_h	12	Number of houses per street in a simulation
N_f	12	Number of dwellings per house in a simulation
N_c	6	Number of connections per street in a simulation

on, we measure the dissimilar distribution of agents in a certain area on the macro level. The most established measure is the index of dissimilarity (ID), which goes back to Duncan and Duncan (1955) and compares the distribution of two groups in a population. We use a modification of the ID, the index of segregation (IS) (Breßler and Harsche 2004), that calculates the distribution of one group in the population in comparison to the total population. This indicator can hold values between 0 (for uniform distribution) and 100 (for maximum separation of the population). For model analysis, we are looking for changes in the ID that occur upon systematic manipulation of the graph while all other parameters are kept constant.

Initialisation

A simulation run is initialised by creating the graph first and then placing the individuals randomly on the dwelling-nodes of the graph.

$$L^{A_i}(0) = \text{random}\{H \mid H \neq L^{A}\}, \qquad (1)$$
$$S^{A}(0) = \text{random}\{G\}.$$

The initial values of the state variables are given in Table 1. They are chosen arbitrary since we are interested in the effects on the segregation index if the structure of the graph is varied. The simulation is initialised many times with systematically varied parameters.

Input

The varied parameters (cp. Table 1) are the tolerance thresholds of the agents, the number of groups, the weightings of the street- and house-nodes, the number of streets, houses, and dwellings in a simulation, and the number of connections of a street node. There are some interrelated parameter settings: For example, if we change the number of streets in a simulation, the total number of dwellings, houses, and agents are also changed because we want to maintain the number of houses per street, the number of dwellings per house, and the ratio of unoccupied dwellings. A simulation is run until there is no movement of agents for 10 time increments. The most important measurement is the segregation index at the end of a simulation run. The simulation is repeated ten times with the same parameter settings to avoid artefacts that result from the random initial placement of individual agents during initialisation.

In detail, the parameters are varied in the following intervals: The tolerance thresholds of the agents are increased from 0 to 1 in increments of 0.01 and decreased from 1 to 0 in increments of 0.01. We use the same values for both tolerance thresholds ($\Psi_{settle} = \Psi_{leave}$, cp. Table 1). The number of groups is varied from 2 to 8, the weightings of the street- and house-nodes are increased from 0 to 1 in increments of 0.01. The number of streets is increased from 5 to 25 in increments of 1, the number of houses is increased from 1 to 25 in increments of 1, the number of dwellings is increased from 1 to 25 in increments of 1 and the number of connections of a single street-node is reduced randomly by one connection until the graph is divided in two unconnected parts. The results of these variations are presented in chapter 3.

Submodels

After initialisation, the agents check their neighbourhood for the ratio $v^{H}(t)$ of agents of the same group to the whole population:

$$v^H(t)=\sum_B \{w_{total} \mid B \in U(A),\ S^B = S^A\} / N^A(t), \qquad (2)$$

where $N^A(t)$ is the number of agents within U(A):

$$N^A(t) = \sum_B \{w_{total} \mid B \in U(A),\ L^B\} \qquad (3)$$

The influence of a cell on another one is specified by w_{total} and is calculated by the product of the weightings that result from the nodes that are located on the shortest path between the two considered nodes H_1 and H_2 (Figure 4):

$$w_{H1,H2} = w_{total} = \prod_{i=1}^{n} w_i \qquad (4)$$

If the influence between two nodes is very small, we define it as nonexistent:

$$if\ \left(w_{Hi,Hj} < 0{,}1\right) \quad then\ \left(w_{Hi,Hj} = w_{total} = 0\right) \quad (5)$$

The size of a neighbourhood results from the weighting of the nodes. Whether or not an agent is satisfied with its current neighbourhood is expressed by its inner state M, which depends on the tolerance threshold Y_{leave}:

$$M^A(t+1) = \begin{cases} 1 & if\ \left(v^H_{old}(t) \geq Y_{leave}\right) \\ 0 & otherwise. \end{cases} \qquad (6)$$

An unsatisfied agent only moves to a new location if its requirements (Y_{settle}) can be fulfilled there as well. Otherwise he stays unsatisfied at its current location. The neighbourhood conditions at a candidate location H_{new} are also evaluated by the equations (2) and (3). The migration rule is:

$$L^A(t+1) = \begin{cases} L^A_{new}(t) & if\ \left(M^A_{new}(t)=1\ and\ v^H_{new}(t) \geq Y_{settle}\right) \\ L^A(t) & otherwise. \end{cases} \qquad (7)$$

An unsatisfied agent can try to change its location only once per time increment.

Simulation experiments

Before we grapple with the question of how the size of a settlement affects the socio-spatial organisation of its population, we will look in the next sections at the conditions for phase transitions in the system, test their robustness, and consider the influence of the density of agents in the system on the segregation dynamics. The definition and identification of critical ranges for the tolerance thresholds of the agents is fundamental for the following experiments.

Critical ranges

In this section, we investigate on which terms phase transitions can be observed. We want to find out which tolerance threshold values result in uncontrolled spontaneous changes in the system. We start with a random configuration of very tolerant agents with $\Psi_{settle} = \Psi_{leave} = 0$. The degree of separation within the population is measured by the IS. To find the critical ranges, the tolerance thresholds are increased incrementally, whereas we always set $\Psi_{settle} = \Psi_{leave}$. Thus the virtual population reduces its tolerance to the point of absolute intolerance ($\Psi_{settle} = \Psi_{leave} = 100$). Similarly, we study the behaviour of the system if the tolerance thresholds are reduced incrementally, meaning that the virtual inhabitants become more and more tolerant.

In Figure 6 the average IS values of ten simulation runs per value setting are shown. The IS values were each measured after the system reached a stable state. The graph above left shows the average IS values when gradually increasing both tolerance thresholds.

Figure 6: Phase transition. As parameter values, we have used $N_G = 3$, $N_s = 12$, $N_h = 12$, $N_f = 12$, $N_c = 6$, $\omega s = 0.5$, $\omega h = 1.0$, $\Psi_{leave} = 0.33$ and $\Psi_{settle} = 0.33$. The number of agents equals 95% of the dwellings. Above left: IS when gradually increasing the tolerance thresholds from 0 to 1. Above right: IS when gradually decreasing the tolerance thresholds from 1 to 0. The graphs below show the move rates (fluctuating line) at the respective tolerance thresholds.

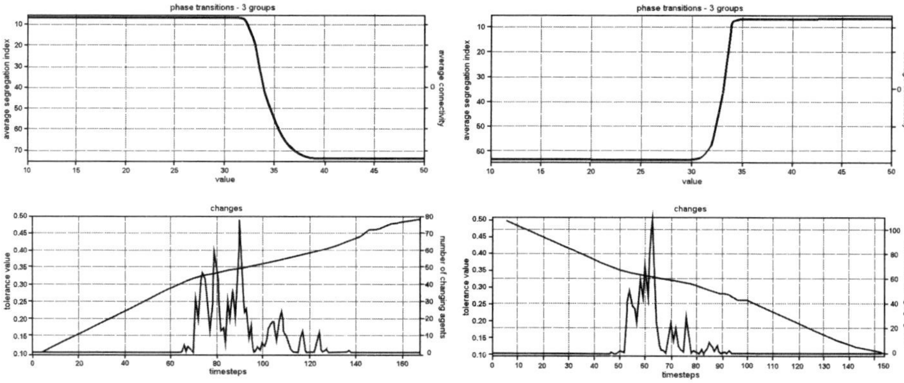

Figure 7: Critical ranges. The different curves show the critical ranges at various numbers of groups N_G in a system. The dash-dotted lines represent the IS values with increasing Ψ values, whereas the continuous lines show the corresponding curves with decreasing Ψ values.

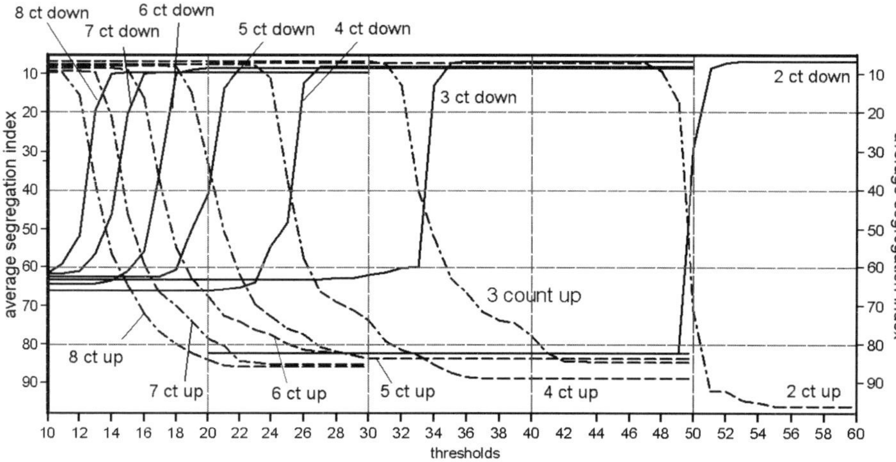

The graph above right shows the IS values when gradually decreasing both tolerance thresholds. As expected, no separation of agent groups occurs at a high tolerance (low thresholds). Anyway it is interesting that no separation of agent groups occurs at very low tolerance (high thresholds), because the high demands of agents (Ψ_{settle}) cannot be fulfilled anywhere. Though the agents are extremely unhappy with their residential situation, they cannot find a satisfying alternative location and therefore cannot improve their situation by moving. Only within a relatively small critical range, which results from the sensitivity analysis of the tolerance threshold (Figure 6 and Figure 7), a phase transition can be triggered

Figure 8: Robustness of the phase transitions. Top: For each curve the number of streets is increased $N_s = 6$, $N_s = 12$ and $N_s = 24$, whereas the total number of houses and dwellings are indirectly increased because they are defined per street and per house ($N_h = 12$, $N_f = 12$). Bottom: For each graph the number of connections per street is increased. In both diagrams, the tolerance thresholds Ψ are assigned to the x-axis and the IS values to the y-axis.

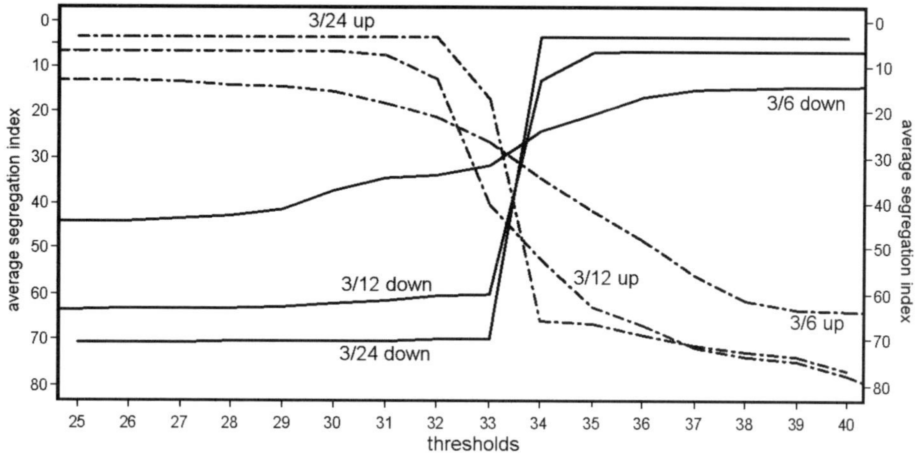

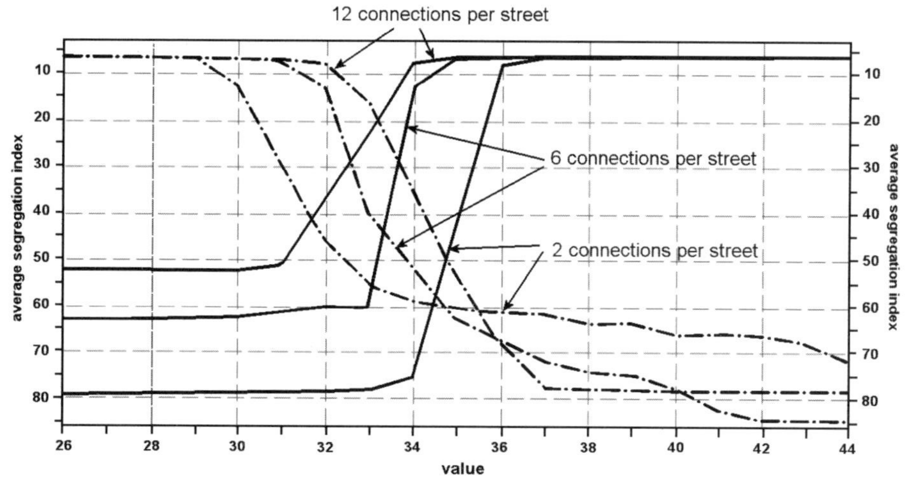

abruptly by a small change of tolerance thresholds.

The results of the analysis shown in Figure 7 make evident that the critical ranges depend on the number of groups N_G in a system. As NG increases, the critical ranges are shifted more to the lower-tolerance thresholds. Considering the Ψ values on the x-axis in Figure 7, where the curves of the increasing and decreasing thresholds intersect, we find that the point of intersection is located almost exactly at $\Psi = 1/N_G$. This might mean that a population is more likely to segregate if it consists of more groups or if the group sizes get smaller.

Finally, we look at the robustness of the critical ranges, firstly, regarding the size of the settlement - which is the number of streets N_s, houses N_h

and dwellings N_f in a system - secondly, the connectivity of the road network, which can be varied by the connections per street, and thirdly, the resident density (agents), which can be expressed by the ratio of occupied to existing dwellings. The comparison of settlements with different sizes in Figure 8 (top) shows essentially no change in critical ranges. Only with very small settlement structures ($N_s = 6$) there is no clear phase transition. The more streets are in a system, the more clear the phase transitions become. The diagram on the right side in Figure 8 shows how the number of connections per street influences the segregation of population. This characteristic value only affects the critical range if we are starting from a settlement without crossroads (graph with two connections per street). In the other

Figure 9: Effect of agent density on segregation. Top: Course of the IS when gradually increasing N_A (n = 190). Bottom: Course of the IS when gradually increasing N_A and with different settlement sizes. The values and the colour code for N_s, N_h and N_f, are stated in the legend. The different lines connect the mean IS values of ten simulation runs per parameter-value setting.

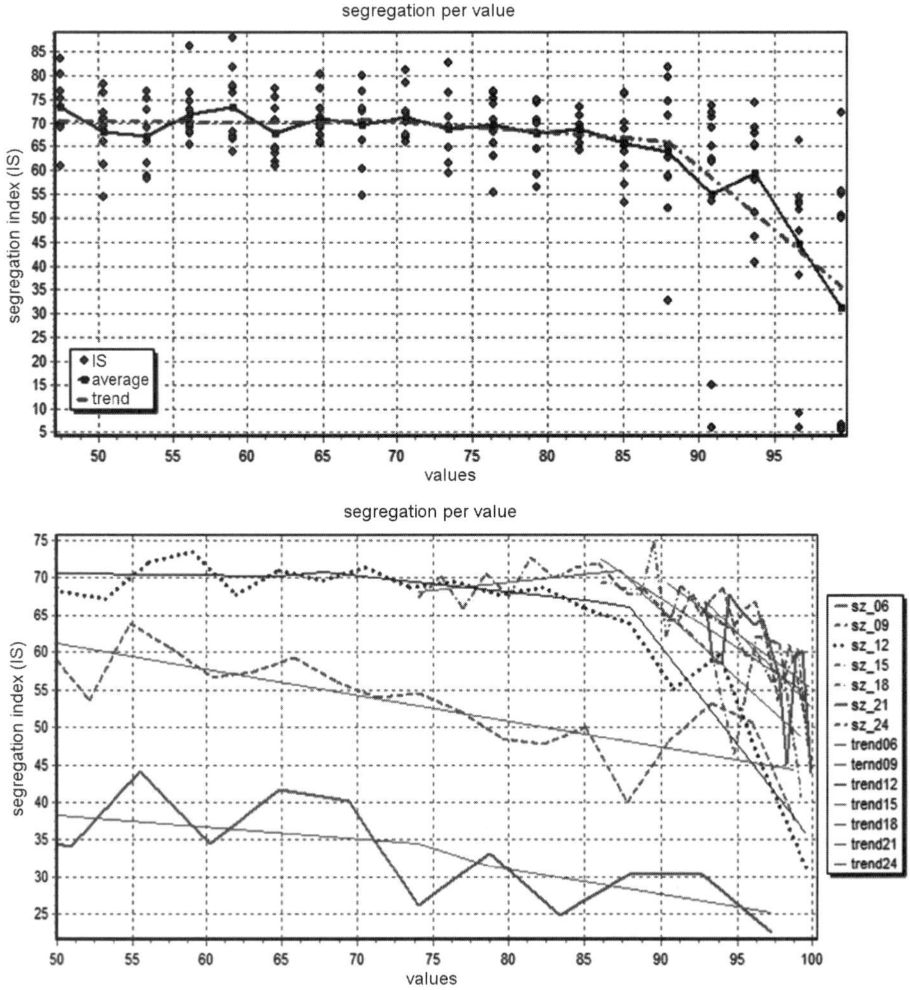

two cases, the course of phase transition is similar.

To investigate the effects of various resident densities, the number of agents N_A is increased gradually by 100 agents, from 48% to 99% of the possible density. The simulation is repeated ten times per setting for NA (total number of simulation runs n = 190). The continuous line in Figure 9 (top) connects the mean IS values that result from ten measurements of the IS.

At an agent density less than 90% the segregation is approximately the same. From a density greater than 90% the IS drops continuously until a clearly lower segregation occur at a density of 99%. This property of the system is based on the effect that increasing density makes it more difficult for unhappy agents to find an unoccupied location where their demand Ψ_{settle} is fulfilled. In other words, we can say that the agents mutually obstruct themselves more often at higher densities than they do at lower densities.

Figure 9 (bottom) shows the same study with different settlement sizes. Mutual obstruction of the agents also starts for most settlements at approx. 90%. The defined default value of 95% occupied dwellings in a system, which is used in the following experiments, seems to be a realistic assumption for prosperous cities (in eastern Germany the vacancy rate is 13–14% and in southern Germany it is about 2% (Sigismund, Waltersbacher et al. 2008)) and allows sufficient room for the agents to move.

Effects of different settlement sizes

With the Circle City model we can now study the effects of different settlement sizes on the socio-spatial organisation of the population that is measured

Figure 10: Variation of the number of streets. Course of IS (ordinate) with increasing N_s (abscissa), N_G = 3, n = 210. Top: Ψ=0.30 (r = -0.11; r^2 = 0.01; s_{-error} = 3.55). Bottom: Ψ = 0.34 (r = 0.53; r^2 = 0.28; s_{-error} = 16.70). The histograms beside the diagrams show the distribution of the data points.

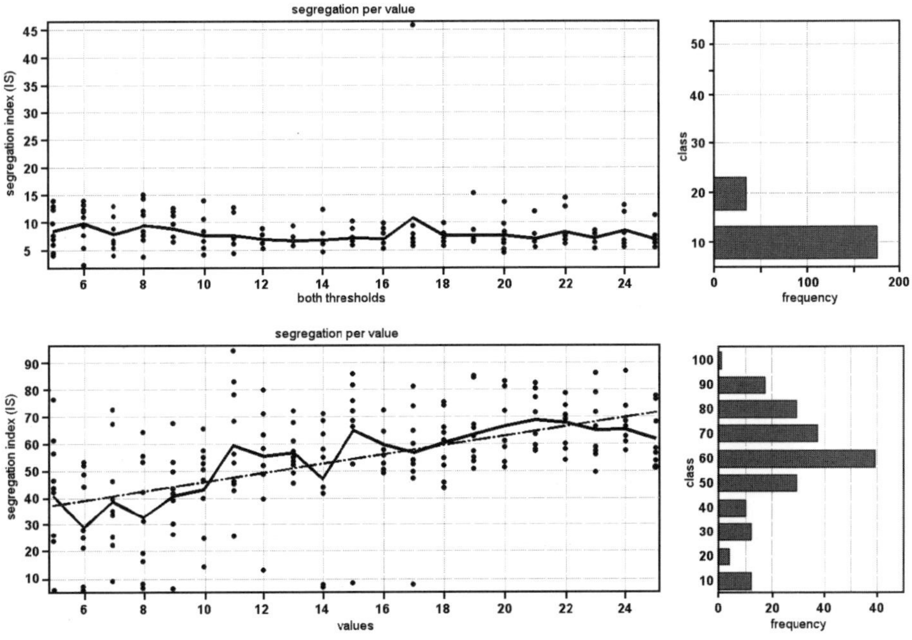

Figure 11: Variation of the number of streets. Course of various characteristic values with increasing number of streets N_s and different tolerance thresholds Ψ (abscissa), $N_G = 3$. Top: Slopes of the regression lines ΔIS (continuous line, left ordinate), standard deviation s_y (dashed line, right ordinate), coefficient of determination r^2 (dotted line, left ordinate), and standard error s_{error} (dash-dotted line, right ordinate). Bottom: The value b_{yx} indicates the intersection of the regression line with the left ordinate (continuous line, left ordinate). The other two curves show the course of the correlation factor r (dashed line, right ordinate) and the IS mean value (dotted line, left ordinate).

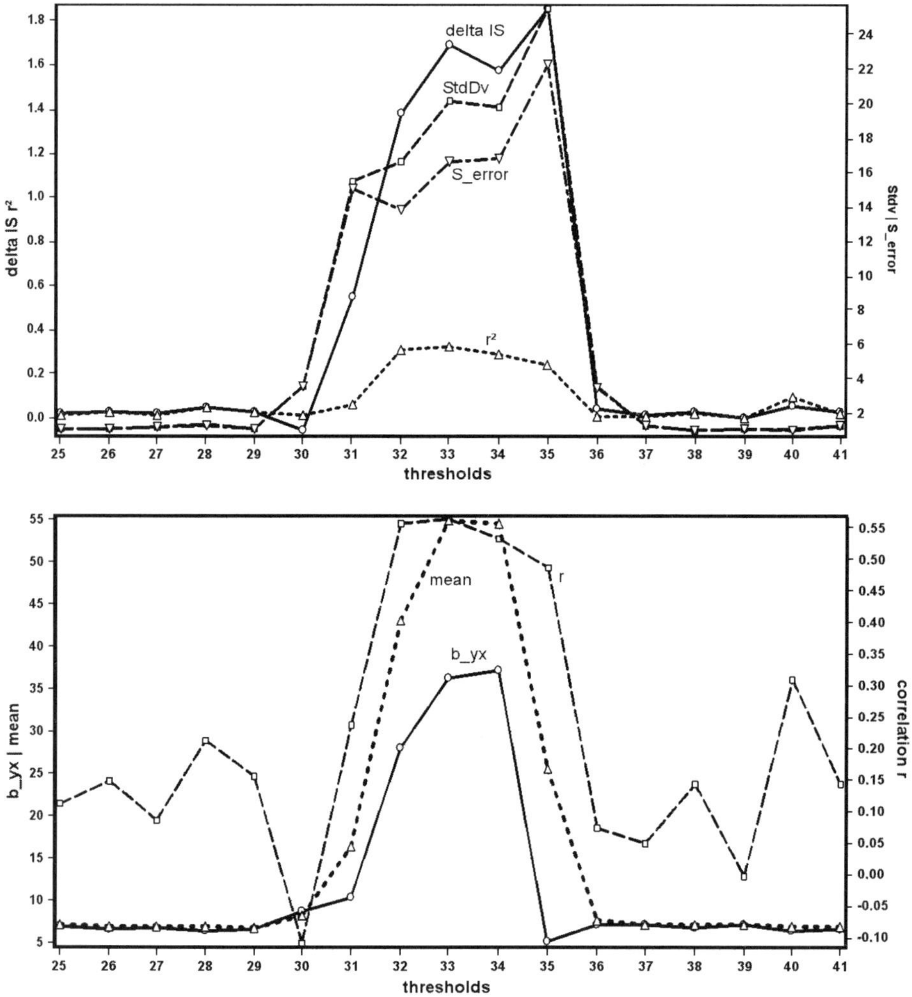

by the IS. For the following examinations, we will particularly concentrate on the critical ranges because the relatively slight effect can probably only be shown there. As specified in Table 1, the default values for the parameter $N_G = 3$ at an agent density of 95%. We vary the values for N_h, N_f and N_c, which describe the size of a settlement. The differences in the segregation of the population that results in dependence on various tolerance thresholds are presented here. For simplification, the thresholds Ψ_{leave} and

Ψsettle are combined and specified with Ψ if *Ψleave* = Ψ_{settle}.

Number of Streets N_s

The analysis of the effects of the number of streets in a system is an important clue for statements about the influence of the settlement size on the population structure. In Figure 10 the course of IS for $N_G = 3$ is shown. On thetop diagram, no effect on the IS of increasing the number of streets from 5 to 25 can be found (Ψ = 0.30 is outside the critical range), whereas on the-bottom diagram, at a slightly higher Ψ value, a clear increase of the IS can be observed (Ψ = 0.34 is inside the critical range). For each value configuration, ten calculations were made (data points, total n = 210) and their mean values connected. The histograms at the right side of the diagrams in Figure 10 are introduced to describe the distribution of the IS values on the y-axis.

For a better graphic presentation of the dependence of the effect of N_s on IS, the individual slopes (ΔIS) of the orange regression line of Figure 10 are calculated for all runs per threshold and marked in the diagram on the left hand side in Figure 11 (continuous line). Both thresholds are increased by increments of 0.01, from Ψ = 0.25 to Ψ = 0.41. For each threshold setting, the number of streets N_s is increased from 5 to 25 in increments of 1 (21 steps) and calculated ten times each. For all other parameters, the system default values are used (Table 1). Therefore, the simulation was run 210 times for each data point in figure 11 (n = 210). We immediately find that the number of streets only has an effect on the IS at tolerance thresholds in the critical range between 0.30 and 0.36. The rest of the Ψ values N_s (and thus, in the figurative sense, the size of a settlement) seems to be insignificant for the separation of the population. For a precise examination of this result, we will use some descriptive statistical characteristic values.

Firstly, we consider the range 0.30 ≥ Ψ ≥ 0.36 (Figure 11, left). Here the slope of the regression line is approximately 0. Therefore it runs horizontally and corresponds to the mean values of the underlying data points. The standard deviation s is relatively small (between 1.0 and 1.3) and thus the dissemination of the data points is very small. The mean value for IS can be read in Figure 11 (bottom). We are primarily interested in the change that results from varying the ratio of the tolerance thresholds to the number of streets. The correlation factor r in this range is very small or 0 (null). Because the regression line has no considerable slope, it is also logical that no linear interrelationship exists between x- and y-values. The standard error s_{error} is nearly s_y.

Next, we consider the range 0.30 < Ψ < 0.36 of the two thresholds. Here a strong increase in the slope of the regression line (continuous line) - up to ΔIS = 1.85 and sy = 25.42 at Ψ = 0.35 - can be noted. The values of the correlation factor increase to r = 0.56. The standard error s error in this range is between 14.89 (97%) at Ψ = 0.31 and 22.21 (87%) at Ψ = 0.35. The graph of the points of intersection of the regression line with the x-axis, b_{xy}, has a course similar to the mean-value curve. These points of intersection show the absolute separation of the population at x = 0 and allow, using the slope of the regression line, the calculation of the absolute separation of the population at all other x-values.

The analysis described until now can thus be interpreted as demonstra-

Figure 12: Variation of the number of houses. Course of various characteristic values with increasing number of houses N_h and different tolerance thresholds Ψ (abscissa), $N_G = 3$. Top: Slopes of the regression lines ΔIS (continuous line, left ordinate), standard deviation s_y (dashed line, right ordinate), coefficient of determination r^2 (dotted line, left ordinate), and standard error s_{error} (dash-dotted line, right ordinate). Bottom: The value b_{yx} indicates the intersection of the regression line with the left ordinate (continuous line, left ordinate). The other two curves show the course of the correlation factor r (dashed line, right ordinate) and the IS mean value (dotted line, left ordinate).

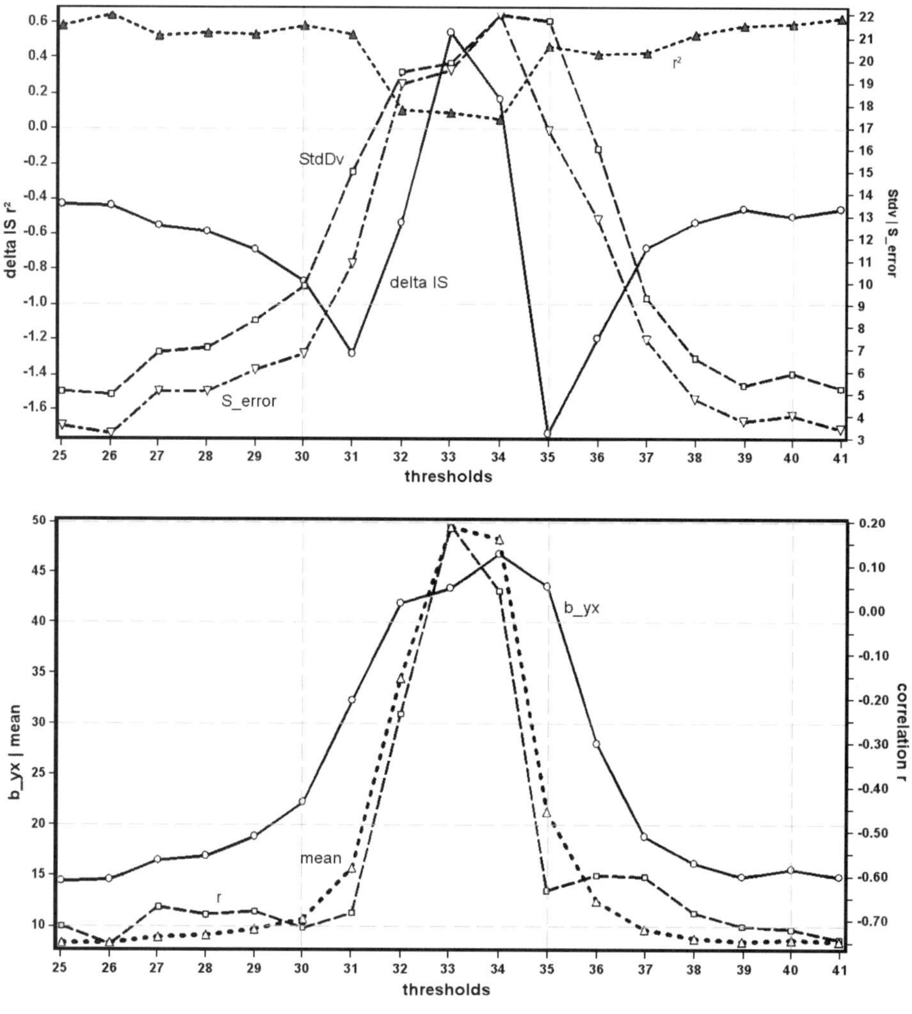

ting that at thresholds in the range 0.30 $\geq \Psi \geq$ 0.36 an increase in the number of streets has no effect on the IS and with that on the separation of the population (cf. Figure 11, top). Since the standard error s error is approximately s_y and reaches a maxim of 1.3, we assume that the size of a settlement (at $N_G = 3$) has no effect on the segregation of the population.

On the other hand, in the range 0.30 $< \Psi <$ 0.36 we can interpret the interrelationship between the number of streets and IS as a strong effect (r = 0.56). The coefficient of determination r^2 shows that a rate of 32% of the variance share at ΔIS can be predicted by the number of streets N_s. This is also reflected in the standard error s error, which is between 87% and 97% of the characteristic deviation. This relatively poor predictability can be attributed partially to the reduction of r b_y some aberrant values (cf. Figure 10, bottom). These mainly result from the initial random distribution of the agents that sometimes lead to configurations in which the agents block each other's attempts to improve their neighbourhood, since the demand Ψsettle for any agent cannot be fulfilled anywhere. In this case, separation fails to develop despite strong unhappiness in the population. This circumstance was already described in the previous with the examination of agent density in a system. In reality this situation can occur if all dwellings in a settlement are occupied and the households do not swap their dwellings.

Due to the required computing time the number of streets cannot be increased as we would like. In particular, many more streets would be required for the representation of a whole city. However, we can assume that the frequency distribution of the IS values in the histograms in Figure 11 (bottom) would just move upward. The effect of the number of streets on the separation of the population within certain ranges of tolerance thresholds Ψ would still be observable.

Although we have only considered the increasing number of streets in this section, the number of houses and dwellings increases automatically with N_s too. The total amount of houses is calculated by

$$N_{\mathrm{h\_ges}} = N_h * Ns\ , \qquad (8)$$

and the total amount of dwellings is calculated by

$$N_{f\_ges} = N_f * N_h * N_s \quad \text{or}$$
$$N_{f\_ges} = N_f * N_{h\_ges}. \qquad (9)$$

This means that for the chosen values, upon increasing the number of streets from 5 to 25, the total number of houses $N_{h\,ges}$ fluctuates between 60 and 300, and the total number of dwellings $N_{f\,ges}$ is between 720 and 3600. For this reason, the described effect cannot only be possibly reduced depeningt on the number of streets, but can also depend on the number of houses or dwellings. Because of this, we investigate N_h and N_f separately in the next two sections.

Number of houses per street N_h

We investigate the effects of the number of houses per street Nh in the same way as we have considered the number of streets. N_h is increased incrementally from 1 to 25 (n = 250) and the slope of each regression line at various tolerance thresholds Ψ is marked as data points in the graph of Figure 12 (top). For the other parameters, the system default values are used (Table 1).

Again, we differentiate between two ranges along the abscissa in the diagrams in Figure 12. Firstly, we consider the thresholds $0.31 \geq \Psi \geq 0.35$. Here the regression lines have negative slopes. Starting from the maximum or minimum Ψ-values on the abscissa of the diagram in Figure 12 (top) the slopes of the regression lines become more negative in the direction of the middle Ψ-values (ΔIS = -0.50 at Ψ = 0.41 and ΔIS = -1.73 at Ψ = 0.35). The standard deviation increases from both sides to the middle (s_y = 5.43 at Ψ = 0.41 and s_y = 23.07 at Ψ = 0.35). The correlation factor r in this range is relatively constant (between r = -0.67 (r^2 = 0.45) at Ψ

= 0.27 and r = -0.76 (r^2 = 0.57) at Ψ = 0.26). The course of the standard error s error follows approximately the course of the standard error s_{error} = 3.46 (64%) at Ψ = 0.41 and s_{error} = 16.47 (71%) at Ψ = 0.35.

Within the second considered range of the thresholds, 0.32 < Ψ < 0.34, the trend of the regression line's slope changes, reaching ΔIS = 0.47 at Ψ = 0.33. However the standard deviation sy = 19.45 is high in contrast to a relatively low r = 0.20 (r^2 = 0.04). The standard error s_{error} = 19.01 (98%) is nearly identical with the standard deviation s_y.

In Figure 13 the distribution of the data points for Ψ = 0.41 and Ψ = 0.33 is shown. Here one can read the absolute change of the IS values for an increasing number of houses per street. The histograms at the right side of the diagrams in figure 13 describe the distribution of the IS values on the y-axis.

Within the first range, 0.31 ≥ Ψ ≥ 0.35, we can assume that the separation of the population decreases with the increase of the houses per street. This negative correlation increases up to r = -0.76 (strong effect). The explained variance reaches a relatively high value of 57%. The small standard error s_{error} = 3.24 (65%) indicates that there are only minor deviations of the IS values from the regression line. However in Figure 13 (top) it is apparent that the method of linear regression analysis is not ideal in this case, since there are to some extent exponential correlations. Nevertheless we can assume that also at high values for N_h, the IS values remain low and the regression line retains a slightly negative slope, even if we cut away the left side of the diagram in Figure 13 (top).

Figure 13: Variation of the number of houses. Course of IS (ordinate) for increasing Nh (abscissa), NG = 3, n = 210. Left: Ψ = 0.41 (r = -0.75; r^2 = 0.56; s-error = 3.46). Right: Ψ = 0.33 (r = 0.18; r^2 = 0.03; s fehler = 19.10). The histograms beside the diagrams show the distribution of the data points.

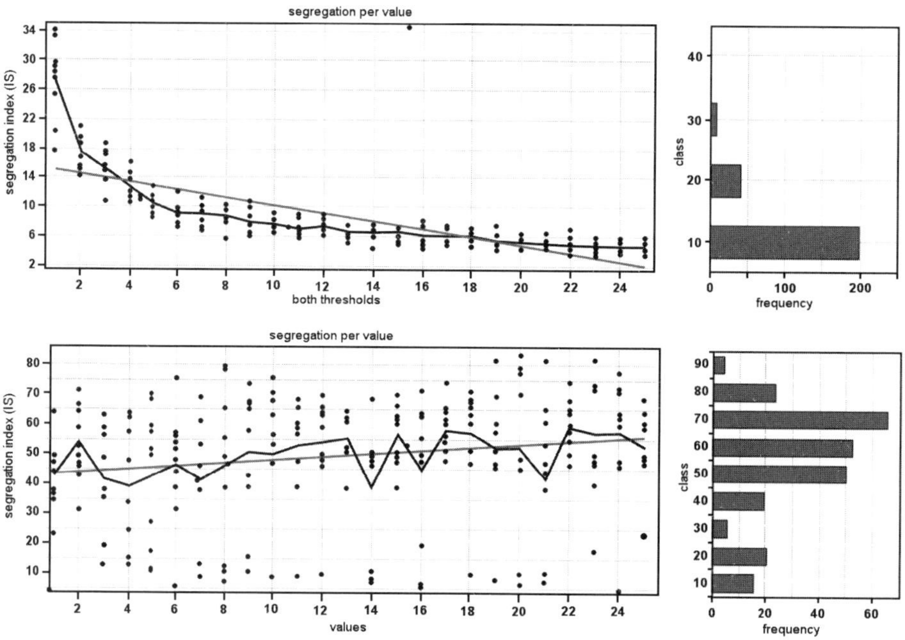

At the second range, $0.32 < \Psi < 0.34$, there is only a weak interre-lationship between the number of houses and the IS, with r = 0.18, which enables almost no prediction with an explained variance of 3% and a standard error s_{error} = 19.10 (98%). However, the distribution of the IS values in the histogram in Figure 13 (bottom) shows that the separation of the population for this Ψ parameter is higher than for the Ψ parameter in Figure 13 (top). It is noteworthy

Figure 14: Variation of the number of dwellings. Course of various characteristic values for increasing number of houses N_h and different tolerance thresholds Ψ (abscissa, values are multiplied by 100), NG = 3. Top: Slopes of the regression lines ΔIS (continuous line, left ordinate), standard deviation sy (dashed line, right ordinate), coefficient of determination r^2 (dotted line, left ordinate), and standard error s_{error} (dash-dotted line, right ordinate). Bottom: The value b_{yx} indicates the intersection of the regression line with the left ordinate (continuous line, left ordinate). The other two curves show the course of the correlation factor r (dashed line, right ordinate) and the IS-mean-value (dotted line, left ordinate).

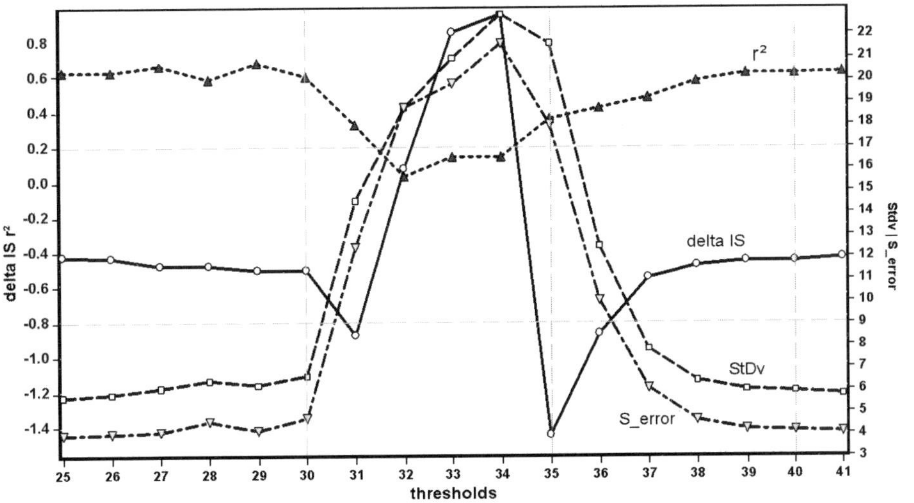

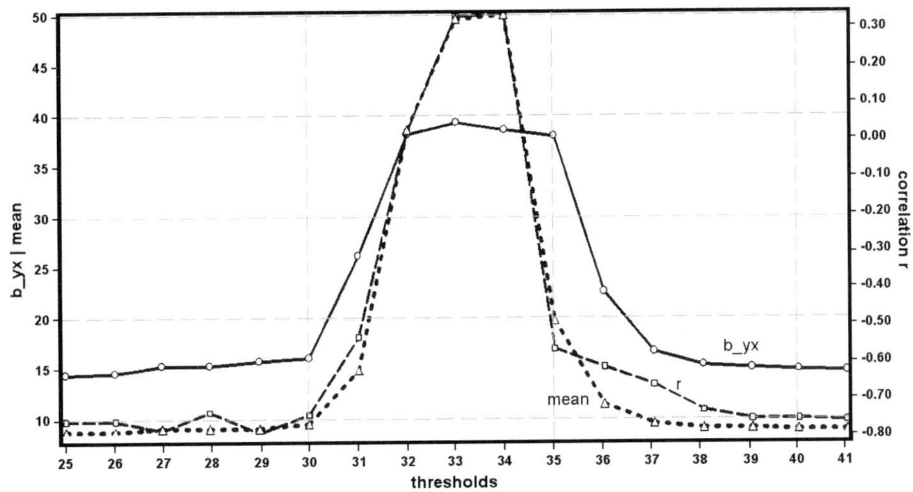

that the correlation between segregation and Ψ values in this small Ψ-value range are reversed in contrast to the Ψ-value range described first (slope of the regression line change from negative to positive). The large difference between the absolute extent of separation within both ranges can be clearly recognised by examining the diagrams in Figure 12, bottom (b_{yx} and IS mean values) and Figure 13. The limited practicability of the linear regression analysis for evaluating the results of these experiments because of the partial exponential correlations was already mentioned. To restrict the statistical part of this investigation we accept this inaccuracy since we primarily want to show certain tendencies and properties of the system instead of carrying out a precise statistical analysis.

Number of dwellings per house Nf

Keeping the systematic of analysis used for the investigation of the number of streets in the last section, we now consider the effects of a systematic increase in the number of dwellings per street N_f from 1 to 25, in increments of 1 (n = 250), on residential segregation. The calculations are repeated ten times for each Ψ value and the mean slopes for each tolerance threshold Ψ are marked as data points in the graph of Figure 14 (top). For the other parameters the system default values are used (Table 1).

Again we differentiate between two ranges on the abscissa at the diagrams in Figure 14. Firstly, we consider the thresholds $0.31 \geq \Psi \geq 0.35$. Here the regression lines have negative slopes. Starting from the maximum or minimum Ψ values on the abscissa of the diagram in Figure 12 (top) the slope of the re-

Figure 15: Variation of the number of houses. Course of IS (ordinate) at increasing N_f (abscissa), $N_G = 3$, n = 210. Left: Ψ = 0.30 (r = 0.75; $r^2 = 0.56$; $s_{error} = 3.63$). Right: Ψ = 0.33 (r = 0.33; $r^2 = 0.11$; $s_{error} = 18.01$). For both diagrams n = 250. The histograms beside the diagrams show the distribution of the data points.

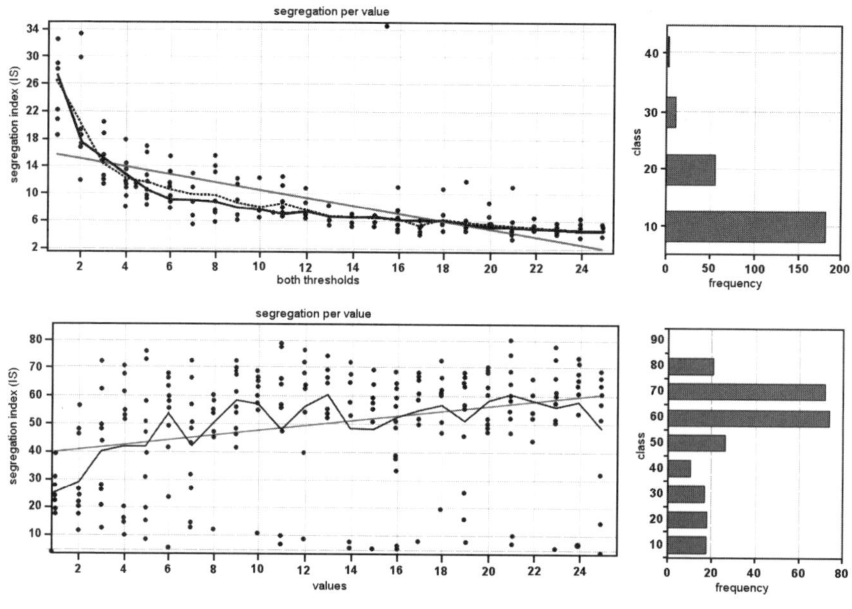

gression lines becomes more negative in the direction of the middle Ψ values (ΔIS = -0.46 at Ψ = 0.41 and ΔIS = -1.44 at Ψ = 0.35). The standard deviation increases from both sides to the middle (s_y = 4.70 at Ψ = 0.41 and s_y = 19.80 at Ψ = 0.35). The correlation factor r in this range stays relatively high (between r = 0.76 (r^2 = 0.58) at Ψ = 0.25 and r = 0.66 (r^2 = 0.44) at Ψ = 0.37). The course of the standard error s_{error} follows approximately the course of the standard error serror = 3.04 (65%) at Ψ = 0.41 and serror = 16.26 (82%) at Ψ = 0.35.

Within the second considered range of the thresholds, 0.32 < Ψ < 0.34, the trend of the regression line's slope changes, reaching ΔIS = 0.91 at Ψ = 0.34. The standard deviation s_y = 19.45 is high at a medium high correlation factor r = 0.33 (r^2 = 0.11). The standard error serror = 18.01 (94%) is nearly identical with the standard deviation s_y.

In Figure 15 the distribution of the data points for Ψ = 0.30 and Ψ = 0.33 is shown. Here one can read the absolute change of the IS values for an increasing number of dwellings per house. The histograms at the right side of the diagrams in figure 15 describe the distribution of the IS values on the y-axis.

Within the first range, 0.31 ≥ Ψ ≥ 0.35, we can assume that the separation of the population decreases with the increase of the dwellings per house. This correlation increases up to r = 0.76 (strong effect). The explained variance reaches a relatively high value of 58%. The small standard error $_{serror}$ = 3.63 (67%) indicates small deviations of the IS values of the regression line (Figure 13, top). Here we also must make reference to the limitation of the linear regression analysis that was already mentioned in the last section.

At the second range, 0.32 < Ψ < 0.34, there is an interrelationship of average intensity between the number of houses and the IS, with r = 0.33. The variance explanation is 11% and the standard error $_{serror}$ = 18.01 (94%) (Figure 15, bottom). It is noteworthy that the correlation between segregation and Ψ values in this small Ψ-value range are reversed in contrast to the Ψ-value range described first (slope of the regression line change from negative to positive). The large difference between the absolute extent of separation within both ranges can be clearly recognised by examining the diagrams in Figure 12nottom (b_{yx} and IS mean values) and Figure 13.

It can be established that results of the investigation about the increase of the number of houses N_h are very similar to the results in this section about the increase of the number of dwellings N_f. By reason of the equation (9), we can assume that the correlation between N_h and IS can be traced back mainly to the fact that an increase of Nh is accompanied by an increase of N_f.

Conclusions and perspectives

Altogether the results of the analysis show that the effects of the built structure of a city on the socio-spatial organisation of its population are relevant if the tolerance thresholds of the population are already in a critical range. A critical range refers to the range of the population's tolerance thresholds within which a small change causes a relatively large effect on the separation of the population. When all population groups have the same size, the critical ranges for the tolerance thresholds are at 1/number of groups (Figure 7).

Within the critical ranges a positive relationship between the size of a city and the separation of the population can be shown by means of the Circle City model. This relationship is enforced if the density of the settlement increases. A higher density can be represented in the model for example with more houses per street or more dwellings per house. With the Circle City model we have demonstrated that firstly, the explicit inclusion of the built structure of a city is an important prerequisite for the simulation of residential segregation and secondly, the size and the density of a settlement on certain conditions can have a relevant effect on the residential segregation.

At future investigations of the Circle City model, various weightings for the connections between streets, houses, and dwellings have to be considered. This would be important to demonstrate the effects that the built design of the streets as well as the arrangement of the buildings have on the separation of the population. Furthermore, it would be interesting to examine if the sociospatial organisation of the population depends on the connectivity of the urban road network. After investigating the principle effects of abstract spatial structures it would be necessary to experiment with representations of real spatial structures and validate the model with data about the residential dynamics of real households.

Acknowledgement

The project was funded by the German Research Foundation (Do 551/18-1).

References

Benenson, I. (1998): Multi-Agent Simulations of Residential Dynamics in the City. In: Computers, Environment and Urban Systems 22(1), pp. 25-42.

Breßler, C. & M. Harsche (2004): Einführung in die Kultur- und Sozialgeographie. Available at: http://www.mygeo.info/skripte/skript_bevoelkerung_siedlung/ (12.11.2006).

Duncan, O. D. & B. Duncan (1955): A Methodological Analysis of Segregation Indexes. In: American Sociological Review 20, pp. 210–217.

Epstein, J. M. & R. Axtell (1996): Growing Artificial Societies: Social Science from the Bottom Up. MIT Press, Cambridge, MA.

Grimm, V., Berger, U. et al. (2006): A standard protocol for describing individual-based and agent-based models. In: Ecological Modelling 198, pp. 115–126.

Grimm, V. & S. F. Railsback (2005): Individual-Based Modeling and Ecology. Princeton University Press, Princeton, NJ.

Portugali, J. (2000): Self-Organization and the City. Springer Series in Synergetics, Berlin Heidelberg.

Schelling, T. (1969): Models of segregation. In: American Economic Review 59, pp. 488-493.

Schelling, T. (1972): A process of residential segregation: neighborhood tipping. In: Pascal A. (ed.) Racial Discrimination in Economic Life. Lexington, MA, Heath: 157-184.

Sigismund, M., M. Waltersbacher, et al. (2008). Housing and real estate markets in Germany 2006. Bonn, Selbstverlag.

Micro-Modelling as a useful tool for planning? The example of modelling gentrification

Rudolf Giffinger und Roman Seidl

Introduction

Over the last decades the understanding of planning changed along with its own experiences and failed impacts steering urban and regional development. Planning approaches had to acknowledge the basic shift from fordist to post-fordist organization of society. These changes very obviously had a variety of impacts on urban and regional development and, therefore, on approaches steering relevant processes. Due to these changes the understanding of plan ning shifted from normative utopian approaches to more incremental and realistic approaches giving up the ambition to steer comprehensively future development through planning in favour of learning based strategies intervening into processes according to specific criteria, i.e. of social justice or territorial cohesion. (Friedman, 1987, Selle, 1996)

Along with these basic structural changes even the possibilities of planning approaches changed: technological progress provided new possibilities of data production, information collection and processing over the last years; the application of GIS for the analysis of spatially referenced information changed the character of information and the speed of decision making even in complex situations. Along with technological progress new types of models – understood as abstraction of complex situations/processes – emerged. Micro-modelling became one of the favoured approaches because of specific characteristics.

In this contribution first we will focus on some characteristics of micro-modelling in front of the changing understanding of the role of planning regarding spatial developments. Its possibilities to understand and steer future development are then elaborated on the example of a rather complex urban phenomenon: gentrification. Based on a discussion of existing modelling approaches we elaborate a micro approach, modelling the specific discussion of the invasion-succession-process (Dangschat, 1988; Dangschat, 1991; Friedrichs, 1996). Concentrating on this topic we try to answer the questions: Which meaning does modelling have in theory and practice? How can we apply the techniques of micro-modelling for planning purposes in a modern understanding of planning.

Modelling as a tool of planning

Basically, any form of spatial planning is concerned with the ambition to understand urban and regional development and to steer future development on different spatial levels in a more or less effective way (Turowski, 2005, p. 893 ff). The understanding of planning and its corresponding methodologies, of course, changed over time whereby this process of modified understanding is not a linear process.

Without describing this process in its manifold aspects (Selle, 1996, p. 36 ff), since many years discussion focuses on the understanding of spatial development, on the role of planning in a complex system and on corresponding approaches with its methods and instruments. Accordingly, Selle (1996, p. 56) elaborated a model of steps and lines which characterizes the evolutionary process of changed planning understanding. In this discussion the main focus remains upon the question how to position planning tasks between the three basic sectors which drive development; i.e. public sector, society and economy. Basically, the understanding shifted from an utopian claim of steering urban and regional development through public planning authorities to an incremental understanding which presumes a less important role of planning – not only in a normative regulatory but in a clear cooperative and moderating role designing spatial processes as one actor besides others. (Selle, 1996, p. 69).

Along with the increasing functional and spatial complexity of spatial processes different forms of governance have been established in the respective spatial context (for instance, metropolitan development, Salet et al., 2003). In particular, cooperation or delegation as specific forms to steer spatial development become increasingly important against the principle of autonomy in face of cross-sectional and cross-border challenges in spatial development.

Along with the strong claim for a moderating role of planners and for processes of participation (stakeholders, residents, NGOs) the discussion raised whether planners should have own visions and stimulating functions as experts or take only a coordinating and mediating role. In front of poor market driven solutions in terms of sustainability and the deterioration of public space, there are good arguments to understand planning profession in a proactive way. This means that even planners have to develop own normative visions from a specific point of view specialized in the integration of complex interests between politics, society and economy in an incremental understanding. Meeting this challenge planners have to be innovative in creating new methods and instruments in order to steer spatial processes as a pro-active partner. This aspect of understanding easily can be described by a statement of John Forester still being of strong importance in recent years: "Professionals are called to profess. However technical it is, planning is an inescapably normative enterprise. Planning schools then must learn how to help students construct, assess, and present normative arguments because planning practice will demand these abilities – not incidentally but fundamentally." (Forester, 1989, p. 201)

Evidently, this understanding is characterized through cooperation between planners and actors and the process is driven from 'practice to critical consciousnesses' and from 'critical consciousness to critical practice'. In this context critical consciousness is result of theory based discussion of visions, and practice is driven through the definition of strategies and implementing actions. This process therefore should be based on 'social learning' – in particular of planners and involved partners (Selle, 1996 after Friedmann, 1987). Modifying the understanding of planning after the so called 'enthusiastic planning phase' around the early decade of the 1970ies we nowadays identify the following characteristics to be important in the recent understanding of planning in an incrementalist perspective (Selle, 1996, p. 290 ff; Healy, 1997; Salet, 2003): cooperative and participatory, analytic process oriented, in different roles (public agenda, local stakeholders, etc.), recursive and network oriented, strategic and project oriented, social learning and innovative.

Modelling

Along with these new understanding of planning old techniques of systemic mostly linear approaches and simple analytical methods had been replaced in order to facilitate and strengthen new forms of social learning based on empirical evidence produced through the application of more sophisticated qualitative and quantitative methods in regional science. Stiens (1996) already showed in his description of methods used in planning that there was a functional shift: tools of empirical analysis were originally used in order to describe and explain and to assess planning measures in a relativly static and closed system/environment. Along with the modified understanding of planning (old and new) methods and models are increasingly applied in order to explore and detect new empirical trends and specific forms of developments. These approaches are increasingly understood as tools which produce new insights between potential individual decisions of stakeholders, relations between different trends and the emergent meaning of such interrelations or decisions on specific aggregate outcomes in future development. Thus, methods and modeling approaches are increasingly used as a tool strengthening social learning and fostering argumentation in corresponding processes which try to steer spatial development. Figure 1 sketches the evolution of prognostic methods and techniques in planning (based on Stiens, 1996, p. 115).

Therefore, along with these changes in the understanding how to plan future developments even the function of research and corresponding modelling techniques changed. Traditional methods and modelling approaches never disappear completely but they have been accompanied and sometimes replaced through new and more complex

Figure 1: Principal methods and techniques of spatial prognosis

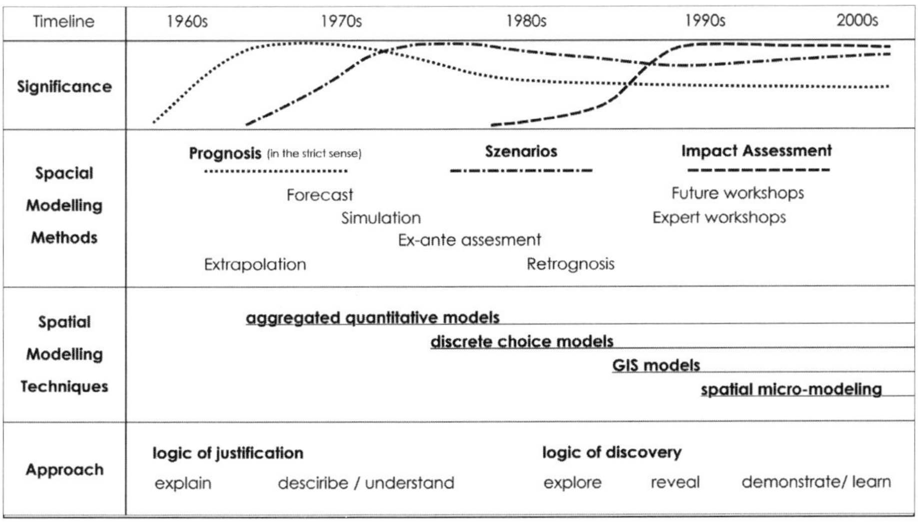

ones. Basically, aggregated quantitative modelling approaches lost importance whereas qualitative methods and in particular GIS based modelling and micro-modelling gained. Within micro-modelling there has been a shift from microsimulation and cellular automatons towards agent based models as these approaches allow to focus on direct interaction between elements rather than just between micro and macro levels (Ballas et al., 2007, p. 2483).

Regarding quantitative approaches there is broad literature on GIS-application and its implications on planning (see for instance Nedovic-Budic, 2000). At the same time there has been a significant boom in spatial modelling. However, approaches of micro-simulation are not very well established regarding spatial planning. Over the last years there seems to be a partial backlash to the use of large scale models (like UrbanSim; Wadell et al., 2008) with all the enthusiasm (Beneson, Torrens, 2005) and strong pitfalls (Lee, 1973, p. 173 ff, Löchl et al., 2007) involved. Considering the disadvantages and trying to apply models within modern incremental planning approaches based on scenarios rather than normative forecasts (Stiens, 1996) we propose a small scale and thus more "bottom-up" approach to modelling.

In order to assess its usability, we will now discuss what micro-modelling means and how this method might be used as a planning tool fostering new pro-active learning based processes of planning.

Micro-Modelling

The starting point for any simulation model is a definition of a subject, e.g., processes and involved objects. In order to elaborate a micro-model relevant theories and observations on the subject are needed which help to define respective structure and relations. In contrast to macro-models the objects are units which are directly observable like inhabitants, dwellings or companies.

One of the main issues in modelling is the level of abstraction. The more extensive a model gets the harder it gets to estimate parameters and to test and validate it. In contrast the results of a simpler model are harder to relate to the modelled subject (Rowe, 1977 calls this the "information paradox"). Thus this decision depends on the aim of the model, the availability of data, etc.

Simulation models are implemented in software. Thus the syntax and semantic of the implementation has to be tested beforehand one is able to test for the validity of the model itself. Doing this the parameter sensitivity must be examined. Finally, under ceteris paribus conditions the examined model is capable to simulate model behaviour under consideration of modified elements or parameters. Having a valid model one can mainly pursue aims: test theory and simulate future developments. Usually testing theory will be done in advance to prognosis.

In social science simulation is mainly used for explanation (Gilbert, Troitzsch, 2005). This helps to explore and formalize processes. Theory which is commonly defined in text is transformed via formal mathematical rules into a simulation-model. In this process disambiguation and contradiction within theory may be found. Finally, modelling provides the comparison of results with the assumptions on model behaviour which may help for testing theory on inner consistency. This becomes particularly important in dynamic models because the effects of feedback and emergence might be hard to predict.

For testing theory an empirical basis of the assumptions on structure, relationship and model parameters is demanded. Given this reliable and valid description model results have to match the results of empirical processes to proof its validity (Weber, 2007). If no such validation can be achieved the model is limited to be perceived as a game (e.g. Game of Life in Gardner, 1970 or Sugarscape in Epstein, Axtell, 1996).

Finally a valid model may be used for the prognosis of a possible future state. Micro-simulation models are explicitly including the effects of feedback or emergence and might model a system out of balance. Thus in contrast to traditional statistical model that exclude such effects the validity of quantitative results might be weaker. Being more focussed on modelling system behaviour than on exact empirical status the results rather predict tendencies and probabilities.

Explicitly modelling complexity allows for understanding and communicating the dynamic effects of regulatory frameworks and interventions. Due to the effects of complexity included into the model the results are limited to short term prognosis because certain interventions or unpredictable events may change a development path. Using micro-models therefore strengthens a predominantly evidence based and learning based understanding of planning.

Modelling Gentrification

In the more than 40 years of gentrification debates there has been a split between supply and demand side approaches and hardly any attempts to combine the both of them (Lees et al., 2008). Within this debate it has often been intended to test theory based on empirical data (e.g. Kecskes, 1996) using traditional forms of modelling. Gentrification processes are complex and can't be modelled well by using macro-models. Using a "bottom-up" approach based on complexity theory and micro-modelling therefore seems to be more appropriate in order to meet the specific characteristics this phenomenon.

Supply-Side-Models

Wu (2003) did a first very basic model that tries to combine the concept of the rent-gap with self-organized-criticality (SOC; Bak, Tang, Wiesenfeld, 1988). He uses a cellular automaton to represent a city landscape. The cells contain a certain amount of population. If the difference between a cell's number of residents and its neighbour's number of residents is below a threshold, a certain number of residents are added to the cell. As this addition may give way to additional differences which lead to further development there is the possibility of avalanches with sizes up to the extent of the whole matrix. Like in a sand pile SOC-model (Bak et al., 1988) he randomly seeds development if no development induced by neighbourhood difference occurs.

The model can be seen as a SOC model as the distribution of the sizes of the avalanches follows a power law distribution (also called 1/f-noise, s. Bak, Tang, Wiesenfeld, 1988; p.364). Wu sees his model as a sketch on how one could model a system that is beyond equilibrium state. Changes in a neighbourhood induce further changes as it can be observed in a gentrification process.

Diappi and Bolchi (2005, 2006, 2008) chose a much more fine grained and thus more realistic approach. Like Wu's SOC-model it is based on a cellular automaton and focuses on differences in development within neighbourhoods. Unlike Wu the cells main attribute is not the population but the capitalized rent. A cell has a potential rent which equals the capitalized rent when the building on the cell is new. The building continually degrades during the simulation and the capitalized rent is decreasing. Thus a gap between the cell's rent and its neighbourhood may arise. In every step there is a set amount of investment to reinvest into buildings having such a gap. So similar to Wu's model certain avalanches may occur but in this model their size is limited by the availability of capital. The model proves that the application of Rent-Gap-Theory combined with the role of investment capital may produce cyclic behaviour.

Demand-Side Models

Unlike Wu or Diappi and Bolchi, O'Sullivan (2002) sketched a demand side model for gentrification. He uses a cellular automaton that represents a certain area within the inner city. The cells contain buildings of a certain quality and inhabitants having a distinct level of income. As the model only covers a part of the inner city people move in and out. Thus gentrification is to be seen when the income level of those moving in is higher than that of the ones leaving. Buildings may be rented or owner -occupied. For both the probability for residents to move rises with the difference in income to the neighbourhood such as the wealthier residents are more likely to leave. If a dwelling is abandoned the price is set to the average of the income in the neighbourhood and its physical condition (which declines over time). The whole area has a status that depends on all the buildings' qualities and all the residents' income levels. New residents are located by drawing a potential resident from a normal distribution whose centre is at the mean of the status of the whole neighbourhood and the cell's price. If the dwelling was for sale and the income of the buyer is above the average of the neighbourhood it will be owner-occupied otherwise it will be let to a tenant. A new owner-occupier will renovate a building if his income is above its physical condition and the buildings condition is below the average of its neighbours.

The model tries to cover several aspects of owner occupation and tenants. Because of this it is very complex for a cellular automaton. On the other hand it does not consider developers. Despite this it seems to be able to show behaviour that has a certain amount of similarity to the situation in the UK. O'Sullivan tries to calibrate it to the developments in Hoxton, London – an area that has a history of gentrification. In this scenario the neighbourhood slowly declines until the share of owner-occupiers falls rapidly. After a short period of accelerated decay reinvestment starts, buildings get renovated and the share of owner-occupiers rises again. So there are longer phases of stability and short phases of decline and reinvestment.

Torrens and Nara (2007) try to model demand side gentrification dynamics by using free agents in cellular Space model (FACS; Portugali, 1999, p.75f.). Agents - representing the population - bid for dwellings with a certain bundle of properties (like accessibility, housing quality, etc.). They are defined by their economic status and preferences for the houses' properties and thus choose them according to their utility in using them (rational-choice). Prices are set to the highest price that can be achieved without the dwelling being vacant for a long time. Thus a hedonic- price model is used (in contrast to the Marxian value-oriented models used by Diappi and Bolchi; O'Sullivan). Due to residents' preferences for the same income and ethnic group in their neighbourhood there is segregation dynamics that lead to gentrification. Beside this process there is no dynamics in the property market – it is merely static. Thus the model seems to be a very complex approach to the demand side which hardly considers emergent effects.

Jackson et al. (2008) also created a FACS demand side model for gentrification. Agents represent the residents of a inner-city location in Boston. They are students, professionals, non-professionals or elderly living in a dwelling represented by cells. The houses' rent levels are assumed to be at the level of the simulated area of Boston. The distance to the university, the Central Business District (CBD) and to shops and restaurants is also a set of properties for every cell. During the simulation the rent levels are constantly increased by a certain factor. In the neighbourhood of professionals this increase is doubled. If an agent is not able to afford the rent he tries to move. If he does not succeed he is removed. Agents prefer places that are either near to the university (students) or the CBD (professionals), where they have lived before and where friends (who are randomly drawn from the same group) live in. Students turn to professionals and non-professionals at a certain rate and professionals and non-professionals to elderly. The elderly eventually die at a certain rate allowing for the influx of new residents.

The model is able to produce a dynamic which is similar to this one observed in the area of Boston. The data and dynamics in Boston was only a four year long relatively linear trend where the share of professionals grew whereas the number of non-professionals and elderly declined. Thus, it could be easy to calibrate even a linear model to show this behaviour. Two aspects of the model seem interesting: It is able to show the effects of the agents' lifecycles and of agents preferring to live near similar friends (which leads to a higher rate of segregation). Beyond these aspects no gentrification dynamics in a self-organized sense seem to exist.

Supply-Demand-Models

Up to now there is only one model by Semboloni (2007) that tries to consider supply and demand. He designs a FACS-model where the cellular space represents the dwellings, the agents are housholds and developers. Houses are defined by housing space (which grows with their distance from the centre), housing quality and rent. The households bid according to their income reduced by their cost to commute and shop. Thus an Alonso-Muth Model (Alonso, 1964, Mills, 1969) is used to represent the trade-offs between price and distance to the centre. Dwellers bid for the houses best according to their individual preferences. In this evaluation space, quality, density, distances and the difference between the agents and the neighbourhoods' income levels are considered. As the agents only have a certain level of tolerance against a neighbourhood with lower average income there is segregation. Calculating tradeoffs between low-density, space and travel-costs leads to a balance between the centre and the suburbs. If a resident moves to a house and is considered to be an owner-occupier and the quality of the dwelling is below a set minimum he will invest to improve its quality. An agent moves by chance or if the ratio between his evaluation of the current situation and the rent is well below that of some friends (chosen by similarity in income and preferences).

Developers are investing into buildings when their quality is below the one of the neighbourhood and the expected rent after the investment is high enough to pay for its cost. The model is tested by growing a city. In the beginning the wealthier households start to settle down in the suburbs. Later on when the inner city is starting to decline there is an invasion of wealthier residents and thus reinvestment. This development is multiplied by developers and thus there is gentrification. Those of the better off who have a high preference for low density and don't travel much remain in the suburbs. Finally there is a rent-bid curve that falls in distance from the centre but rises again in the suburbs. Semboloni is able to show that in his model gentrification is dependent on two factors: owner-occupiers who invest at the beginning of an invasion and tolerance by these pioneers. If there is either no tolerance or no private investment there is no gentrification.

Invasion-Succession-Models

Invasion-Succession

Dangschat (1988) proposed a phase-model for gentrification that describes the change in population in a gentrifying area: the double invasion-succession-cycle. In 1996 Friedrichs criticised this model in that it could not be applied to empirical data and hoped that this could be solved by combining it with a tipping-model. Thus a tipping-model for the double invasion-succession-cycle will be proposed.

The invasion-succession-cycle is a concept applied on the tradition of the Chicago School (Burgess et al., 1925). The model describes a process that can be seen in expanding cities. Working class population invades areas inhabited by the middle class. As this becomes visible those better off start to sell their houses as they are expecting a decline in house value. These houses are then occupied by the poorer population (succession).

This change is typically not proceeding continually but has a point where it starts to accelerate – the "tipping point" (Schelling, 1978, p. 101). This is due to the fact that people base their decisions (to move in or out) on observation of e.g. the share of new population or renovated houses. Empirical studies on the invasion-succession cycle found tipping points to be within a share of 5 to 25% (Friedrichs, 1996, p. 24).

In the double invasion-succession-cycle (Figure 2; based on Dangschat, 1988, p. 281) the population is first displaced by pioneers. They have a high level of tolerance towards the primary population. A higher social and possibly also financial capital allows for their success in the housing market and thus invasion.

As a consequence of the pioneers' invasion the neighbourhood is getting more attractive for people with higher economic status. Gentrifiers are moving in and due to their ability to pay higher rents than both of the former groups they displace both of them.

Invasion-Succession-Automaton

This process will be modelled by using a cellular automaton whose cells can

Figure 2: Double invasion-succession-cycle (Dangschat, 1988; p. 281)

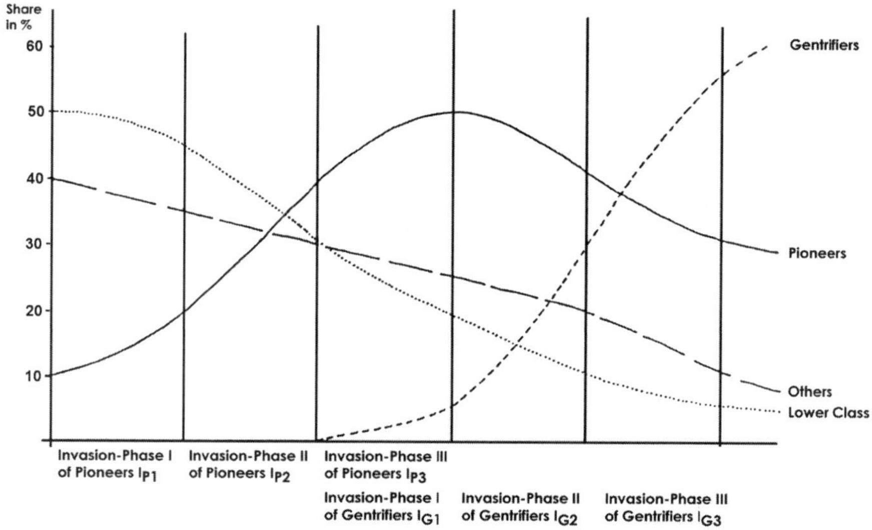

have different possible states to represent the three groups of population. The model was implemented in NetLogo (Wilenksy, 2006).

In our simulation the automaton is a rectangular matrix that is 50 by 50 fields wide and forms a torus. The three groups in the cycle (initial population, pioneers, gentrifiers) are represented by the cell belonging to one of three possible groups (a, b, c). The Invasion will be driven by levels of tolerance towards the share of members of the other two groups (called o and p below) in a Moore-neighbourhood.

If there are too many members of the other groups nearby the resident will leave. Thus the basics of the model are similar to Schelling's (1978, p. 173 ff.) segregation model.

$$Q_i = \{g_i, T_i\}$$
$$T_i = \{t_{oi}, t_{pi}\}$$

Q_i ... *state of cell i*
g_i ... *group of cell i* $g \in \{a, b, c\}$
T_i ... *tolerance of cell towards other groups*
t_{oi} ... *tolerance of cell i towards other group o*
t_{pi} ... *tolerance of cell i towards other group p*

Whether a cell is content with its neighbours is calculated by considering the share of its neighbours belonging to the other two groups and its tolerance levels towards these groups:

$$\begin{Bmatrix} \frac{n_o}{t_{oi}} & if\ t_{oi} < 1 \\ 0 & otherwise \end{Bmatrix} + \begin{Bmatrix} \frac{n_p}{t_{pi}} & if\ t_{pi} < 1 \\ 0 & otherwise \end{Bmatrix} \leq n_n$$

n_o ... *number of members of other group o in the neighbourhood*
t_{oi} ... *tolerance towards other group o*
n_p ... *number of members of other group p in the neighbourhood*
t_{pi} ... *tolerance towards other group p*
n_n ... *number of cells in the neighbourhood (in the case of a moore neighbourhood 8)*

If a cell is vacant there is a certain probability for it to be occupied by a group. This change possibly makes others in the neighbourhood unhappy as their share declines. So they will leave too - which may trigger further changes. Therefore the system's state is increasingly critical as the invading groups grow. This criticality is rooted in the basic dynamic effects of the rules. Thus the model is designed

Figure 3: Flow of the invasion-succession automaton

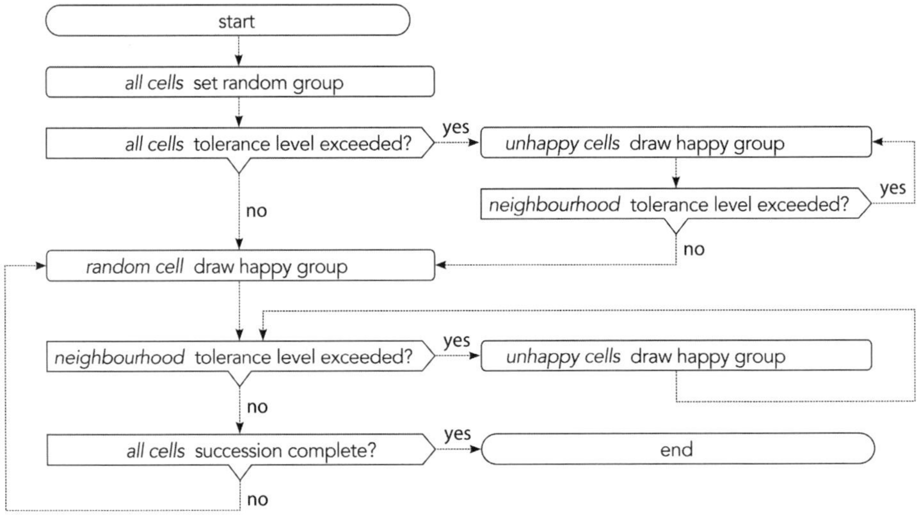

to show self-organized criticality. The sand – in the sense of the sand pile SOC models (Bak et al., 1988) – is applied by randomly changing a cell's resident to a member of another group – presumed that he will be happy with his neighbours. Then the neighbours are asked if they are happy. If they aren't they leave and are replaced by members of the other group. Then their neighbours are asked and so on. This leads to avalanches whose sizes are observed to determine their size distribution.

Behaviour

The initial state of the model is set to be occupied only by members of the first group (a). To allow for invasion the tolerance of the second group (b) towards the first one thus must be 100%. To allow for the second invasion of the last group (c) their tolerance towards b is also 100%.

The last groups' tolerance towards the first one is set to 0% to limit the possibilities for their invasion to areas where the second group has already dislocated all of the first group. Similar to tipping points in tipping models (e.g. Schelling, 1978, p. 173 ff.) all other tolerances in the basic test setup are set to 75%.

The model shows behaviour (s. Figure 5) similar to that of the double invasion-succession-cycle (s. Figure 2). The second (b) groups share increases until the last (c) one begins to grow. Finally c successes and dislocates all the others.

Figure 4: Tolerance levels between the groups

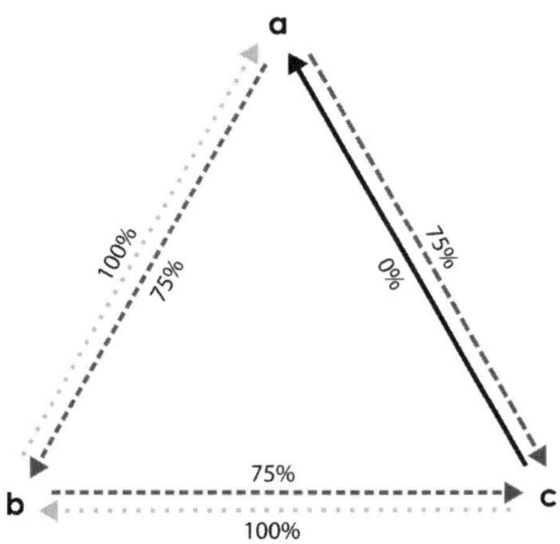

Figure 5: Share of the groups over simulated time

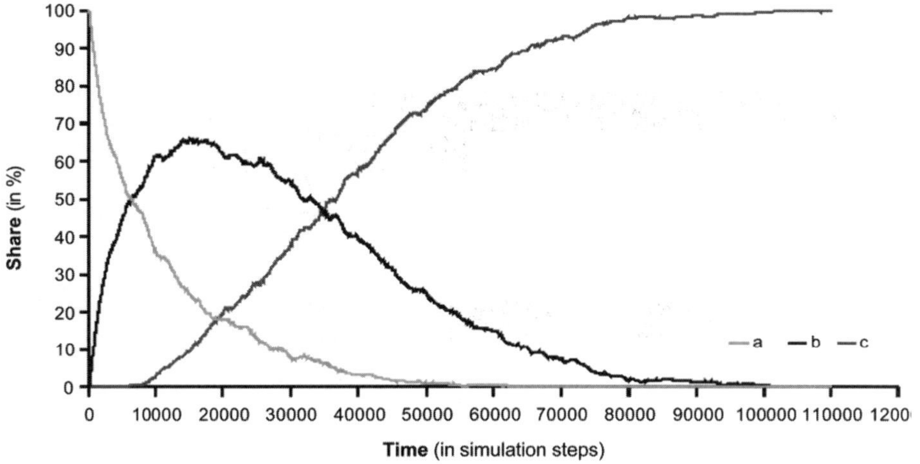

Self-Organized-Criticality

Systems governed by self-organized-criticality show a power law distribution in the size of the changes induced by a single change in a time step of the simulation (Bak et al., 1988). Thus these were examined by testing the goodness of fit towards a generic power law distribution that does not

$$p(x) = \frac{\propto -1}{x_{min}} \left(\frac{x}{x_{min}}\right)^{-\propto}$$

$p(x)$ ... ***propability density for x***
$\propto$ ··· ***parameter of the power law function***
x_{min} ... ***minimum for x considered***

consider small events (below xmin):

The parameters were estimated by using maximum-likelihood method and tested with a Kolomogorow-Smirnow-Test. The Procedure follows the one proposed by Clauset et al. (2009) and used matlab-scripts delevoped by Aaron Clauset (2009).

The maximum-likelihood test for estimated parameters of the standard scenario (xmin=2, α=4.73) gives a p value of 0.14. Within common assumptions for this test the sizes can thus be seen to be power law distributed but the goodness of fit is not very good.

As Figure 6 indicates, small events are less and large ones are more probable than in the estimated power law distribution. So the differences between the automaton and the original sandpile models (Bak et al., 1988) are evident.

Figure 6: Frequency of the number of changes per time step

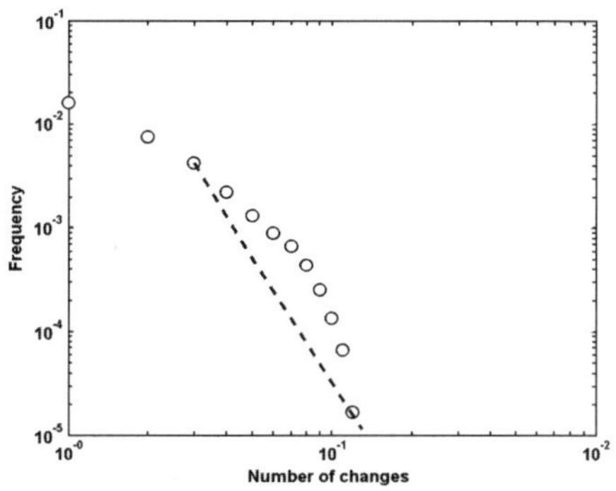

Figure 7: Number of simulation steps until complete succession

b-c	c-a \ a-c:	0					25					50					75				
	a-b:	0	25	50	75	100	0	25	50	75	100	0	25	50	75	100	0	25	50	75	100
0	0	-	-	-	-	-	-	-	-	-	-	-	-	-	-	-	-	-	-	-	-
	25	4	100	907	1928	1631	4	100	988	2533	-	4	-	-	-	-	-	-	-	-	-
	50	4	100	617	876	876	4	100	673	876	2368	1071	12857	-	-	-	36874	-	-	-	-
	75	4	100	218	218	218	4	100	218	218	900	4013	6798	15347	13639	21005	65993	84167	167120	-	-
	100	4	5	5	5	5	6	106	145	145	145	2463	3702	3858	5134	6549	66489	63720	60813	91646	91646
25	0	4	128	-	-	-	4	128	-	-	-	4	128	-	-	-	4	128	-	-	-
	25	4	100	1444	2278	2749	4	100	1281	2873	6066	4	100	-	-	-	4	100	-	-	-
	50	4	100	748	876	876	4	100	852	1361	1459	4	2859	7049	8285	31413	4	36202	-	-	-
	75	4	100	218	218	218	4	100	484	537	762	1801	3024	4329	4084	8011	40080	31437	65097	59361	67095
	100	4	5	5	5	5	4	76	81	81	215	2249	2590	3680	2929	4735	23937	36857	60321	66024	63491
50	0	3285	3751	7347	32128	-	3285	3751	7347	35357	-	3285	3751	6943	49295	-	3285	3751	6943	-	-
	25	3193	3878	5703	9704	10683	3193	4112	6036	10170	51509	3193	2867	5908	13129	-	3193	2867	5852	-	-
	50	2809	1879	1528	2995	3276	2656	1781	1336	3893	9692	2656	3077	3618	7128	20865	2656	1931	21917	42123	-
	75	651	704	536	422	218	760	525	1070	1992	6292	536	1248	4760	4435	10746	442	9681	17280	40060	-
	100	4	1439	166	5	5	193	76	1953	562	2438	256	1150	4068	7251	10900	12183	8125	21904	31579	113531
75	0	55475	60469	69538	99544	-	55475	60469	69538	84975	-	55475	60469	62156	81406	-	55475	60469	62156	110011	-
	25	79318	65183	67238	63502	91248	79318	54466	61835	79040	-	79318	75720	61042	72297	-	79318	75720	61042	86671	-
	50	53363	64673	62672	72195	55495	49919	86517	58184	72048	66040	49919	56427	76102	67378	197669	49919	67184	53046	63392	-
	75	67878	59536	57296	50447	45382	67640	45929	62578	62110	66606	50788	70728	66206	93829	143904	69155	52311	55001	69573	-
	100	63009	56177	49010	49346	52074	50951	59475	54002	65549	67716	43340	55658	51396	125281	-	59903	55566	67293	96578	-

Parameter Sensitivity

To test for parameter sensitivity the tolerances of the first (a) group towards the others, the tolerance of the second (b) towards the third (c) and the one of the last group towards the first group were varied. Figure 7 shows the results of this variation by the number of simulation steps until the last group has invaded the whole area. The cases with a tolerance of 100% of the first group towards the last group and the second group towards the last group are not shown as they never lead to complete succession of the last group.

Figure 7 indicates that complete succession is faster with smaller tolerances towards the last group (c). Also a small level of tolerance of the first group towards the second one speeds up the process. If the last group is intolerant towards the first one but the first group is tolerant towards the last one the succession slows down or does not succeed. This effect is even stronger with high levels of intolerance of the second group towards the last one or the first one towards the second one as this weakens the moderating role of the second group.

Figure 8: Maximum share of the yellow group by varied parameters

b-c	c-a \ a-c, a-b	0: 0	0: 25	0: 50	0: 75	0: 100	25: 0	25: 25	25: 50	25: 75	25: 100	50: 0	50: 25	50: 50	50: 75	50: 100	75: 0	75: 25	75: 50	75: 75	75: 100
0	0	-	-	-	-	-	-	-	-	-	-	-	-	-	-	-	-	-	-	-	-
	25	0	2,04	14,96	26,41	22,95	0	2,04	16,99	31,37	-	0	-	-	-	-	-	-	-	-	-
	50	0	2,04	10,34	13,96	13,96	0	2,04	10,73	13,96	26,72	0	4,23	-	-	-	0	-	-	-	-
	75	0	2,04	3,81	3,81	3,81	0	2,04	3,81	3,81	11,73	0	3,88	13,49	15,26	16,07	0	5,23	14,65	-	-
	100	0	0,04	0,04	0,04	0,04	0	1,46	1,88	1,88	1,88	0	2,73	3,69	3,69	3,65	0	2,92	4	4	4
25	0	0	98,35	-	-	-	0	98,35	-	-	-	0	98,35	-	-	-	0	98,35	-	-	-
	25	0	2,04	27,3	29,26	33,33	0	2,04	21,07	34,76	44,14	0	2,04	-	-	-	0	2,04	-	-	-
	50	0	2,04	12,19	13,96	13,96	0	2,04	13,8	20,68	19,88	0	2,96	20,72	27,53	36,56	0	2,35	-	-	-
	75	0	2,04	3,81	3,81	3,81	0	2,04	8,23	8,69	11,96	0	4,23	17,38	18,69	22,65	0	5,15	17,65	22,72	24,6
	100	0	0,04	0,04	0,04	0,04	0	0,96	1,08	1,08	2,77	0	2,35	9,69	9,69	13,53	0	2,77	11,11	13,84	14,4
50	0	94,96	98,35	85,39	61,36	-	94,96	98,35	85,39	61,98	-	94,96	98,35	83,66	61,55	-	94,96	98,35	83,66	-	-
	25	84,7	93,58	83,04	50,52	48,98	84,7	92,62	85,85	47,37	49,79	84,7	92,93	80,39	47,02	-	84,7	92,93	79,58	-	-
	50	62,94	61,25	23,53	26,95	32,76	64,9	60,52	19,65	29,72	39,29	64,9	57,52	28,95	35,76	39,98	64,9	60,94	28,64	40,6	-
	75	2,65	2,04	3,81	3,81	3,81	3	2,04	9,3	19,34	27,64	0,92	2,92	23,88	26,57	28,76	1,38	6,42	28,18	29,3	-
	100	0	0,46	0,92	0,04	0,04	0,88	0,96	3,54	9,42	14,38	0,46	2,35	14,76	17,07	21,88	0,46	5,84	17,57	21,88	24,9
75	0	94,96	98,35	86,54	66,63	-	94,96	98,35	86,54	64,05	-	94,96	98,35	84,78	65,44	-	94,96	98,35	84,78	66,01	-
	25	84,58	93,58	81,81	61,4	60,17	84,58	92,62	83,12	61,48	-	84,58	92,93	79,82	55,44	-	84,58	92,93	79,82	52,44	-
	50	73,89	71,2	67,09	57,82	59,55	73,01	72,47	60,05	58,36	46,25	73,01	70,28	59,17	51,44	41,06	73,01	72,36	57,21	44,56	-
	75	57,55	51,63	53,48	51,83	51,86	56,98	46,33	51,36	50,33	44,52	57,4	48,06	51,83	47,56	34,14	57,25	47,6	51,13	40,41	-
	100	45,91	46,44	44,68	45,79	45,14	44,87	44,64	45,87	45,66	37,18	45,02	45,62	43,29	43,06	-	45,02	45,44	45,83	36,76	-

Figure 8 shows the maximum share of the second group (b) dependent on the variation of the levels of tolerance. Higher levels of tolerance of the second group towards the succeeding last group (c) delay the last group's succession and thus lead to a higher maximum share of the second group (b). In contrast a higher level of tolerance of the last group (c) accelerates their growth and thus reduces the maximum share of the second one (b). The effect of tolerance between a and b varies according to the tolerance between b and c. If the level of tolerance of b towards c is below 50% the maximum share of b grows with higher levels of tolerance of a towards b. This is because of more stable mixed neighbourhoods of a and b helping to prevent the succession of c.

If the tolerance of b towards c is above 50% this effect is reversed. Lower levels of tolerance of a towards b lead to an acceleration of the displacement of a by b. Because of the higher tolerance of b towards c there is no need for mixed neighbourhoods of a and b to prevent b's immediate displacement.

With a level of tolerance of b towards c of 50% the effect of a change in tolerance of a towards b on b's maximum share depends on the tolerance of a towards c.

Results and Outlook

According to the testing of the change-size distributions the model is characterized by self-organized criticality. It also resembles the assumed progression of the double invasion succession-cycle. The assumptions of the model obviously provide for such model behaviour.

The parameters regarding tolerance are not based on empirical data. Thus the model is only able to show the consistency of the assumptions in their dynamic consequence. Also the neighbourhood definition and the use of tolerance limits are quite arbitrary choices but could easily be replaced by others without changing the basic model behaviour. Thus the model is a suitable approach in order to describe and simulate the dynamics of a double invasion-succession cycle.

The proposed model is based on the idea to use limits of tolerance for a tipping-process (e.g. also used by Schelling, 1978). Like Wu's (2003) supply-model it uses the concept of a sand pile automaton and of self-organized criticality. Thus, this approach is able to model a system out of balance. But in contrast to Wu's approach it is a demand-model. The use of tipping-models for a demand model therefore proves to be an interesting approach.

Like all demand-models for gentrification it is able to show a process of upgrading only, but no cyclical behaviour of disinvestment and reinvestment. Therefore, it would be an interesting perspective to combine this model with more powerful approaches like the one elaborated by Semboloni (2007). Semboloni's demand-supply model also considers a level of tolerance - against poorer neighbourhoods. Semboloni draws incomes from a Pareto-distribution. Segregation is then driven by the agents' bids on dwellings depending on their evaluation of neighbourhood. Agents mainly move randomly allowing for no direct effect of neighbourhood change. Thus his model resembles a situation where tenants are granted long-time contracts and their rents don't change but due to general inflation.

Semboloni does not explicitly model pioneers. As he draws tolerance from a normal distribution there are at least some agents with a high level of tolerance who act as pioneers by triggering invasion into a low-status neighbourhood. In contrast to common assumptions about pioneers and gentrifiers there is no negative correlation between tolerance and income. Thus it would be interesting to modify Semboloni's model to allow for the investigation of the mediation by pioneers and dislocation due to rent-increase because of improvement of the neighbourhood status as it can be seen in the model proposed in this paper.

Conclusions

Based on the very fundamental patterns of gentrification and using the advantages of micro-simulation we elaborated a simple approach on how to integrate the concepts of self organised criticality and tipping models in order to simulate a duplicate invasion-succession cycle. However, the model presented above is not based or tested on empirical data.

Thus it is a game-like proof of concept. It is able to proof that the assumptions allow for a dynamic that resembles the assumed behaviour. Like all demand-side simulation models for gentrification, basically, it is able to show a process of population change based on neighbourhood effects only, but no cyclical behaviour of disinvestment and reinvestment.

Two perspectives of further development are obvious: (1) a further elaboration of an integrated demand-supply model trying to calibrate their behaviour based on empirical data. Anyhow, calibration might be hard to achieve because of the specific characteristics of micro-models; (2) an elaboration of the effects of policy which would not necessarily need for a calibrated model. In a more didactic approach a corresponding model could try to illustrate the effects of rent laws, rent control or urban renewal policies on housing market dynamics and gentrification processes.

However, which conclusions can we draw regarding micro-simulation as a new tool of planning?

The concept of small scale models elaborated in this paper differs from concepts of traditional quantitative large scale modelling. In opposite to the latter, micro-simulation usually focuses on a more heuristic approach considering even qualitative factors of influence. However, when it comes to the use of models in planning it is important to be able to discuss the results (Winkelmann, 1998, p. 63f.). In using models to communicate and learn this must not be limited to modelling experts. The small scale approach can allow easily comprehending and thus discussing its assumptions and implementation.

Besides, there might be a path that leads to usable larger scale models. As the development of spatial micro-models progresses new patterns may emerge. If these patterns are published for others (as with the UrbanSim initiative) at least experts from the field of spatial modelling may discuss their results critically. This might make them more reasonable than traditional large scale approaches which are hard to understand (Wadell et al., 2008).

Models which explicitly include the effects of complexity do not produce quantitative prognosis in the traditional sense but provide new knowledge on distinct paths of spatial development and their respective probabilities. This form of knowledge production matches and strengthens the current paradigm in planning theory to think of planning as an ongoing process of steering development based on scenarios of possible futures and the identification of strategic options (Stiens, 1998, p. 128 f). In this context planning is more concerned with securing the availability of future options than with determining the current optimal choice (Scholles, Mühl, 2001, p. 200).

Models like the one proposed above are able to test theory for their implicit consequences. Thus, they should foster information and communication in strategic approaches. For instance, including aspects of housing market or land regulation into such type of model strategic planning would be able to go beyond a pure communicative paradigm. Rules, regulations and other means of power that seem to be invisible in the communicative planning approach (Reuter, 2004, p. 57f.) can be made explicit in their meaning and, thus, can be discussed directly. In this understanding the role of planners would not revert to a "god-father" model of planning where the planner hides behind political normative provisions seeing himself as an engineer merely calculating optimal decisions. Thus, micro-simulation provides a proactive understanding of planning in front of the ongoing change of regime from government to governance (Haus, Heinelt, 2004, p. 167ff.).

References

Alonso, W. (1964): Location and land use. Havard University Press, Cambridge.

Bak, P., Tang, C. & K. Wiesenfeld (1988): Self-organized criticality. In: Physical Review A, 38(1), pp. 364-374.

Ballas, D., Kingston, R., Stillwell, J. & J. Jin, (2007): Building a spatial microsimulation-based planning support system for local policy making. In: Environment and Planning A 39, pp. 2482-2499.

Beneson, I. & P. M. Torrens (2004): Geosimulation – Automata-based modeling of urban phenomena. John Wiley & Sons, Chichester UK.

Burgess, E. W., Park, R. E. & R. D. McKenzie (1925): The City. University of Chicago Press, Chicago.

Clauset, A., Shalizi, C. R. & M. E. J. Newman (2009): Power-law distributions in empirical data. In: SIAM Review 51(4), pp. 661-703.

Clauset, A. (2009): Power-law Distributions in Empirical Data, available at: http://tuvalu.santafe.edu/~aaronc/powerlaws/ (5.12.2009).

Dangschat, J. S. (1988): Gentrification: Der Wandel innnenstadtnaher Wohnviertel. In: Friedrichs, J. (Hrsg.): Soziologische Stadtforschung, Sonderheft 29 der Kölner Zeitschrift für Soziologie und Sozialpsychologie, pp. 272-292.

Dangschat, J. S. (1991): Gentrification - Indikator und Folge globaler ökonomischer Umgestaltungen, des Sozialen Wandels, politischer Handlungen und von Verschiebungen auf dem Wohnungsmarkt in innenstadtnahen Wohngebieten. (unveröffentlichte Habilitationsschrift), Hamburg.

Diappi, L. & P. Bolchi (2005): Investments, Rent Fluctuations and Gentrification Waves in the Inner-City: an Interpretative Multi-Agent Model. Paper at CUPUM 2005, London.

Diappi, L. & P. Bolchi (2006): Gentrification Waves in the Inner-City of Milan: A multi agent/cellular automata model based on Smith's Rent Gap theory. In: van Leeuwen, J. P. & H. J. P. Timmermans (eds.): Innovations in Design & Decision Support Systems in Architecture and Urban Planning. Springer, pp. 187-201.

Diappi, L. & P. Bolchi (2008): Smith's rent gap theory and local real estate dynamics: A multi-agent model. In: Computers, Environment and Urban Systems 32, pp. 6-18.

Epstein, J. M. & R. Axtell (1996): Growing Artificial Societies - Social Science from the Bottom Up. The Brookings Institution, Washington D.C.

Forester, J. F. (1989): Planning in the Face of Power. University of California Press, Berkeley.

Friedman, J. (1987): Planning in the Public Domain. From Knowledge to Action. Princeton, New Jersey.

Friedrichs, J. (1996): Gentrification: Forschungsgegenstand und methodologische Probleme. In: Friedrichs, J. & R. Kecskes (Hrsg.): Gentrification - Theorie und Forschungsergebnisse. Leske + Budrich, Opladen, pp. 13-41.

Gardner, M. (1970): The fantastic combinations of John Conway's new solitaire game „life". In: Scientific American 223, pp. 120-123.

Gilbert, N. & K. G. Troitzsch (2005): Simulation for the Social Scientist. Open University Press, Maidenhead.

Haus, M. & H. Heinelt (2004): Politikwissenschaftliche Perspektiven auf den Stand der Planungstheorie. In: Altrock, U., Günther, S., Huning, S. & D. Peters (Hrsg.): Perspektiven der Planugnstheorie. Leue, Berlin, pp. 167-186.

Healy, P. (1997): Collaborative Planning: Shaping Places in Fragmented Societies. Macmillan, London, Basingstoke.

Jackson, J., Forest, B. & R. Sengupta (2008): Agent-Based Simulation of Urban Residential Dynamics and Land Rent Change in a Gentrifying Area of Boston. In: Transactions in GIS 12(4), pp. 475-491.

Kecskes, R. (1996): Ein Simulationsmodell des Gentrification-Prozesses. In: Friedrichs, J. & R. Kecskes (Hrsg.): Gentrification - Theorie und Forschungsergebnisse. Leske + Budrich, Opladen, pp.

Lee, D. B. (1973): Requiem for Large-Scale Models. In: Journal of the American Planning Association 39(3), pp. 163 – 178.

Lees, L., Slater, T. & E. Wyly (2008): Gentrification. Routledge, New York.

Löchl, M., Bürgle, M. & K. W. Axhausen (2007): Implementierung des integrierten Flächennut-zungsmodells UrbanSim für den Grossraum Zürich - Ein Erfahrungsbericht. In: disP 168, pp. 13-25.

Muth, R. F. (1969): Cities and housing: the spatial pattern of urban residential land use. University of Chicago Press, Chicago.

Nedović-Budić, Z. (2000): Geographic Information Science Implications for Urban and Regional Planning. In: Journal of Urban and Regional Information Systems Association 12(2), pp. 81-93.

O'Sullivan, D. (2002): Towards micro-scale spatial modelling of gentrification. In: Journal of Geographical Systems 4, pp. 251-274.

Portugali, J. (1999): Self-organization and the city. Springer, Berlin.

Reuter, W. (2004): Planung und Macht – Positionen im theoretischen Diskurs und ein pragmatisches Modell von Planung. In: Altrock, U., Günther, S., Huning, S. & D. Peters (Hrsg.): Perspektiven der Planungstheorie. Leue, Berlin, pp. 57-78.

Rowe, W. D. (1977): An anatomy of risk. John Wiley & Sons, New York.

Salet, W., Thornley, A. & A. Kreukels (2003): Metropolitan Governance and Spatial Planning. Spon Press, London.

Schelling, T. C. (1978): Micromotives and macrobehavior. W.W. Norton & Company, New York.

Selle, K. (1996): Was ist bloß mit der Planung los? Erkundungen auf dem Weg zum kooperativen Handeln. In: IRPUD (ed.) Dortmunder Beiträge zur Raumplanung, Bd. 69, 2. Auflage, Dortmund, pp.

Semboloni, F. (2007): Gentrification as an emerging phenomenon from a real estate market. Congress of the European Regional Science Association (47th Congress), PARIS - August 29th - September 2nd, 2007.

Scholles, F. & G. Mühl (2001): Quantitative Prognosemethoden. In: Scholles, F. & G. Mühl (Hrsg.): Handbuch der Theorien + Methoden der Raum- und Umweltplanung, Dortmunder Vertrieb für Bau- und Planungsliteratur, Dortmund, pp. 199-203.

Stiens, G. (1996): Prognostik in der Geographie. Westermann, Braunschweig.

Stiens, G. (1998): Prognosen und Szenarien in der räumlichen Planung. In: Akademie für Raumforschung und Landesplanung (ARL), Methoden und Instrumente räumlicher Planung, Hannover, pp. 113-145.

Torrens, P. & A. Nara (2007): Modeling gentrification dynamics: a hybrid approach. In: Computers, Environment and Urban Systems 31, pp. 337-361.

Turowski, G. (2005): Raumplanung (Gesamtplanung). In: Akademie für Raumforschung und Landesplanung (Hrsg.): Handwörterbuch der Raumordnung. P. 893 – 898, Hannover.

Waddell, P., Wang, L. & X. Liu (2008): UrbanSim: An Evolving Planning Support System for Evolving Communities. In: Brail, R. (ed.): Planning Support Systems for Cities and Regions. Lincoln Institute for Land Policy, Cambridge, MA, pp. 103-138.

Weber, K. (2007): Simulationen in den Sozialwissenschaften. In: Journal for General Philosophy of Science, Heft 38, pp. 111-126.

Wilensky, U. (2006): NetLogo, available at: http://ccl.northwestern.edu/netlogo/

Winkelmann, U. (1998): Modelle als Instrumente der räumlichen Planung. In: Akademie für Raumforschung und Landesplanung (ARL), Methoden und Instrumente räumlicher Planung, Hannover, pp. 51-66.

Wu, F. (2003): Simulating Temporal Fluctuations of Real Estate Development in a Cellular Automata City. In: Transactions in GIS 7(3), pp. 193-210.

Simulating informal urban growth in Dar es Salaam, Tanzania – A CA-based land-use simulation model supporting strategic urban planning

Alexandra Hill & Christian Lindner

Introduction

One of the most fundamental and most radical changes recently experienced by humanity is global urbanisation. At the beginning of the 21st century humanity entered the 'century of cities' (Kofi Annan) with the majority of people worldwide living in urban areas. In the last decades new so called megacities (cities with more than 5 or 8 or 10 million inhabitants; depending on definition) emerged in Asia, Latin America and Africa (Kraas & Mertins 2008; Bronger 2004; Fuchs et al. 1994; Mertins 1992). The UN highlight that "the developing world, where megacities will be the hallmark of future urbanisation, is experiencing extraordinary spurts of growth in its largest cities" (UN 2007). This contribution focuses on informal urban growth processes in Dar es Salaam, one of the most rapidly growing large cities in Africa. The city's urban poor cannot afford to participate in the formal land and housing markets while local planning authorities due to constraints in financial resources and personnel capacity are barely able to respond to the enormous urban growth, including the ensuing demand for new residential plots. Thus, most of the growth is absorbed by informally developed settlements mushrooming at the peri-urban edges of the city (Hill & Lindner 2009, 2010; Kreibich et al. 2008; Kombe & Kreibich 2006; A. Lupala 2002; J. Lupala 2002). Urban informality has been recognised as a predominant "mode of urbanisation" (Roy 2005) in many developing countries, and the discourse on informality has shifted from a formal-informal dichotomy (de Soto 2000, Hall and Pfeiffer 2000) to a formal-informal continuum (Roy & AlSayyad 2004). This is a consequence of the fact that the formal and informal sectors are strongly interlinked and complement each other. Although research has focused on informal urbanisation and many studies on megacities were conducted in the past years knowledge about the determinants and mechanisms influencing informal urban dynamics is still incomplete. As a step towards a better understanding of the underlying processes this study aims to explore, model and simulate informal urban growth processes using the modelling approach of cellular automata (CA). CA models seek to reproduce development patterns and the underlying mechanisms on an aggregate level. Based on the knowledge generated, they can be used to forecast future trends and demonstrate the likely impacts of selected planning measures. A particular focus of this contribution is to demonstrate the potential of transport infrastructure as an access point for planning intervention to guide future informal development.

CA-based modelling

The CA-based modelling approach aims at reproducing the outcomes of underlying micromotives as macro-behaviour (Schelling 2006) by means of an aggregated, cell-based simulation model on land-use transformation. Most cities are characterised by a permanent transformation reflecting demographic, economic or technological changes. "Almost all cities are undergoing continual growth, change, decline, and restructuring - usually simultaneously" (White & Engelen 1993: 1176). Moreover, cities show the typical features of complex systems as they exhibit several of the characteristics of complexity: fractal dimensionality, self-similarity, self-organisation, emergence and non-linearity (Barredo et al. 2003; Torrens 2000; Torrens & O'Sullivan 2001). "Studying cities as complex systems the critical point is that the sum of the parts corresponds to urban morphology whilst the whole is represented by urban pattern dynamics. From this perspective, dynamics become the keyword in order to study cities from a holistic perspective" (Barredo et al. 2003: 149) in accordance to von Bertalanffy's notion that "the whole is greater than the sum of the parts" (1968: 18).

In general, an automaton is "a processing mechanism; a discrete entity, which has some form of input and output and internal states" (Benenson & Torrens 2005: 349). Automata are discrete with respect to time, space and cell states and the state of the automaton in a subsequent time step is determined by a set of rules. CA models add a spatial component by combining a number of automata in the form of a lattice of cells whereas every cell represents a single automaton. The classical CA characteristics can be described as follows: "At the most rudimentary level, a cellular automaton is an array or lattice of regular spaces or cells. At any given time, a particular cell is in one of a finite number of allowed states, and that state will change according to the states of the neighboring cells in the lattice according to a uniformly applied set of transition rules. Cells alter their states iteratively and synchronously through the repeated application of these rules. A CA is thus composed of four principle elements: a lattice, a set of allowed states, neighborhoods defined by the lattice, and transition rules. In addition (...) a fifth, temporal component" might be considered (Torrens & O'Sullivan 2001: 163). In the context of urban applications the states often represent land uses. CA basically reproduce urban dynamics on a local-scale level assuming decentralised neighbourhood diffusion processes (like influence of existing land uses in a defined neighbourhood area) as one of the main determinants of urban dynamics. But many urban CA in the geographical context are not CA strictly speaking as the classical definition of CA "has been loosened by practice" (Ménard & Marceau 2005: 693). Rather they are considered to be CA-based models because to make standard CA suitable for urban simulations significant changes are necessary. There seems to be no "defining characteristics of standard CA that researchers, in their effort to make CA-based models more realistic, have not been able to discard" (Couclelis 1997: 167). These modifications aim to make the CA approach more suitable to reproduce urban phenomena by for instance including factors such as land-use suitability, accessibility, zoning designations, constraint areas or a perturbation factor.

Although CA models are by definition very simple and base on a set of straightforward rules they can produce surprisingly complex forms (Barredo et al. 2003; Torrens 2000; White et al. 1999; Couclelis 1988; May 1976). Despite their simplicity some classes of CA are capable of 'universal computation' (Wolfram 1984a, 1984b, 1988) which implies that "some types of CA can mimic or reproduce behaviours with high level of complexity, such as of physical, biological or social complex systems" (Barredo et al. 2003: 150). Consequently, the descriptions of complex systems need not be themselves complex, let alone 'complicated' (Couclelis 1988; May 1976). But any modification of a CA model must be approached with care due to the fact that "CA exist, as it were, on the edge of chaos. Launching enthusiastically into dramatic modification of models, without a full appreciation of the system dynamic implications, may produce simulations that we do not fully understand" (Torrens & O'Sullivan 2001: 165). Accordingly, the model conception should be held as simple as possible, but at the same time precise and powerful enough as to illustrate the functioning principles of the system under study.

CA-based models are highly suitable for modelling land-use development in environments that are characterised as being data-poor and experience rapid urban growth and informal development. As the CA modelling paradigm is largely oriented towards simplicity such models do not require a highly sophisticated database - a feature which is of significant advantage considering the application in data-poor environments typically found in developing countries. Incorporating neighbourhood effects CA intrinsically allow for reproducing dynamic and complex processes, particularly those that are strongly influenced by local-scale factors. Urban growth in an informal environment is neither guided by strategic planning nor zoning specifications. "Urban planners and managers have increasingly found themselves confronted by new spatial forms and processes, the drivers of which often lie outside the control of local government" (UN-Habitat 2009: xxiii). In sub-Saharan Africa, informal urban development is strongly influenced by bottom-up, small-scale processes largely relying on social networks.

Integrated land-use transport modelling has a forty years long tradition (Wegener 2004) with CA-based applications for the urban context emerging mainly since the 1990s (see for instance White & Engelen 1994; Batty & Xie 1994; Batty 1997). Although informal urban growth processes in developing countries seem to be particularly suitable to be reproduced as diffusive processes using CA-based approaches, surprisingly, so far only few examples applying CA-based models in developing countries in sub-Saharan Africa and their rapidly growing cities exist such as those for Lagos, Nigeria (Barredo et al. 2004), or for Yaoundé, Cameroon (Sietchiping 2004).

Modelling informal residential dynamics in Dar es Salaam

Dar es Salaam is Tanzania's largest city in terms of population and one of the fastest growing urban agglomerations in Africa. Like in most rapidly growing cities of the developing world informal residential growth constitutes a major component of urban change (see Figure 1). With projected urban population growth rates of about 4% between 2010 and 2025 Dar es Salaam is expected to become a megacity (i.e., a city with more than 5 million inhabitants) before 2020 (UN 2010). Rapid urban growth under poverty has outstripped the capacities of planning authorities to cope with the enormous pace of urban expansion. As a consequence informal settlements absorb almost all homeseekers leading to rapid urban sprawl into the unplanned periphery and the emergence of informal sub-centres in peri-urban areas. The urban structure of Dar es Salaam resembles a four finger pattern following the four major arterial roads. Informal urban sprawl in Dar es Salaam can be described as a diffusion process progressing in wave-like concentric rings from the city centre towards the periphery of the city in combination with broadening ribbons following the arterial roads (see Figure 1).

Despite intensive densification and consolidation processes in existing settlements, most of the urban population growth is accommodated on formerly vacant or agricultural land. The urbanised area of Dar es Salaam has rapidly expanded during the past decades. Inside this research‘s study area (see Figue 2) between 1982 and 2002, more than 15,500 ha of formerly vacant or agricultural land have been transformed into urban land uses. About 77% of these new developments can be classified as informal settlements, whereas only roughly 15% were transformed into planned residential areas. Furthermore, about 8% of the newly developed areas accommodate other urban land uses, e.g., areas for industrial or commercial uses, transport or public services (see Figure 3). High land consumption mainly results from the fact that single storey buildings prevail and affordable large plots can only be found at the urban fringes. Although land consumption is high, there is still much vacant land available at the urban fringe. This is one specific characteristic which distinguishes Dar es Salaam and other large cities in sub-Saharan Africa from many rapidly growing cities in other parts of the developing world. The availability of land contributes to increasing rates of land consumption, as well as to very low settlement densities at the outer fringes of the city.

The impact of formal planning in Dar es Salaam is very limited. The latest masterplan approved dates back to 1979. In 2002, the Government of Tanzania introduced the so called ‘20,000 Plots Project’ which aimed at reducing the shortage of planned, surveyed and serviced plots in the city by means of creating new self-sufficient, self-sustained satellite settlements in selected peri-urban areas of Dar es Salaam. At the same time the project was meant to control land speculation and corruption (URP 2006: 3 f). Contrary to official statements the project did not focus on the urban poor but rather continued to exclude them - and women in particular - from the formal land market (Kombe 2009). The project sites have been selected according to availability largely far apart from existing urban services and informal sub-centres in the urban periphery (Figure 2). Moreover, the delivery of plots through the 20,000 Plots Project only satisfies a fraction of the overall demand.

In Dar es Salaam, estimates assume that approximately 70% of the population live in informally developed settlements with severe deficits in basic services and infrastructure supply

Figure 1: Land-use development in Dar es Salaam in 1982, 1992 and 2002

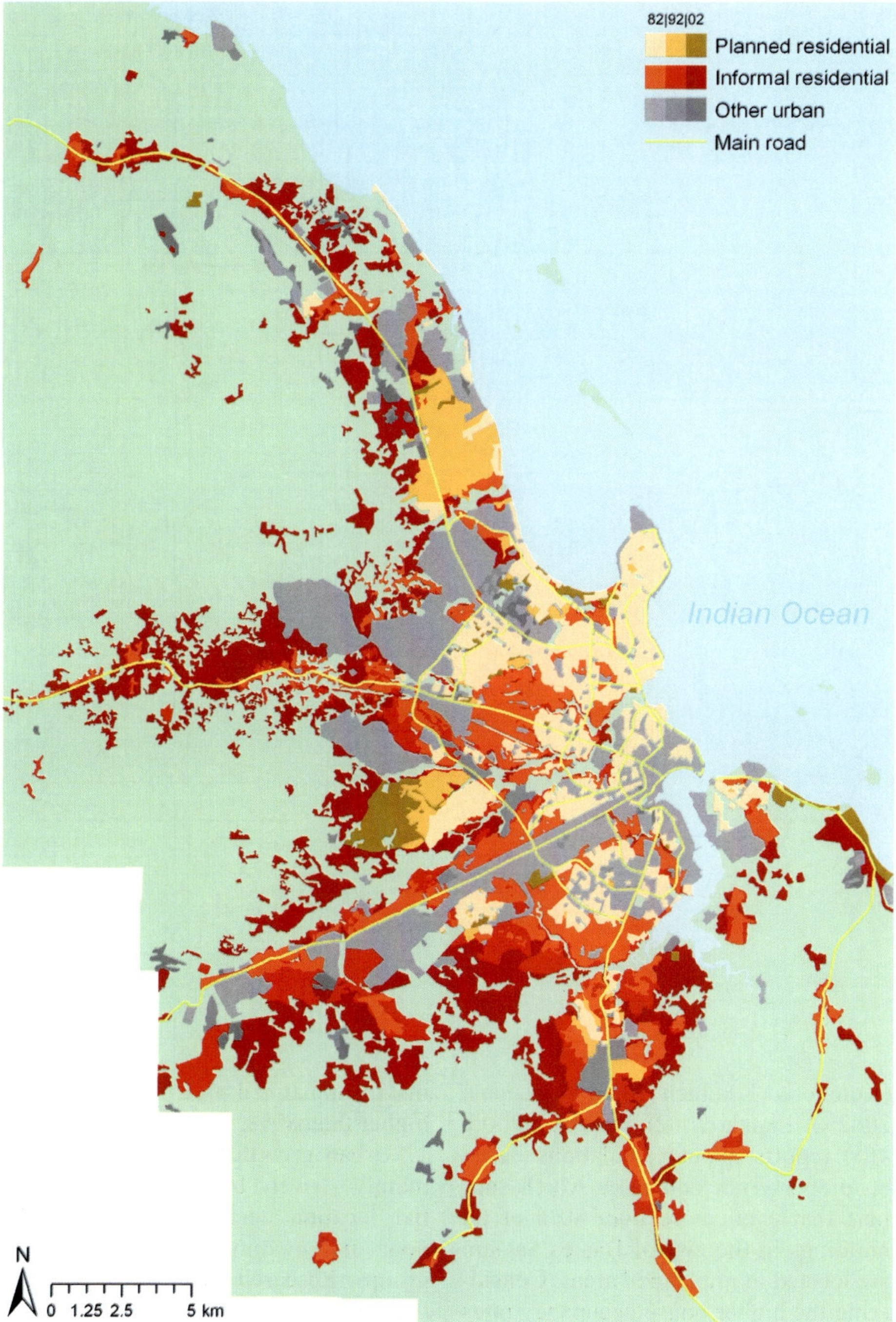

Source: Own illustration; based on land-use and transport network data provided by ITC, Enschede; updated for 2002 by IRPUD, Dortmund

Figure 2: Location of 20,000 Plots Project areas and informal peri-urban sub-centres

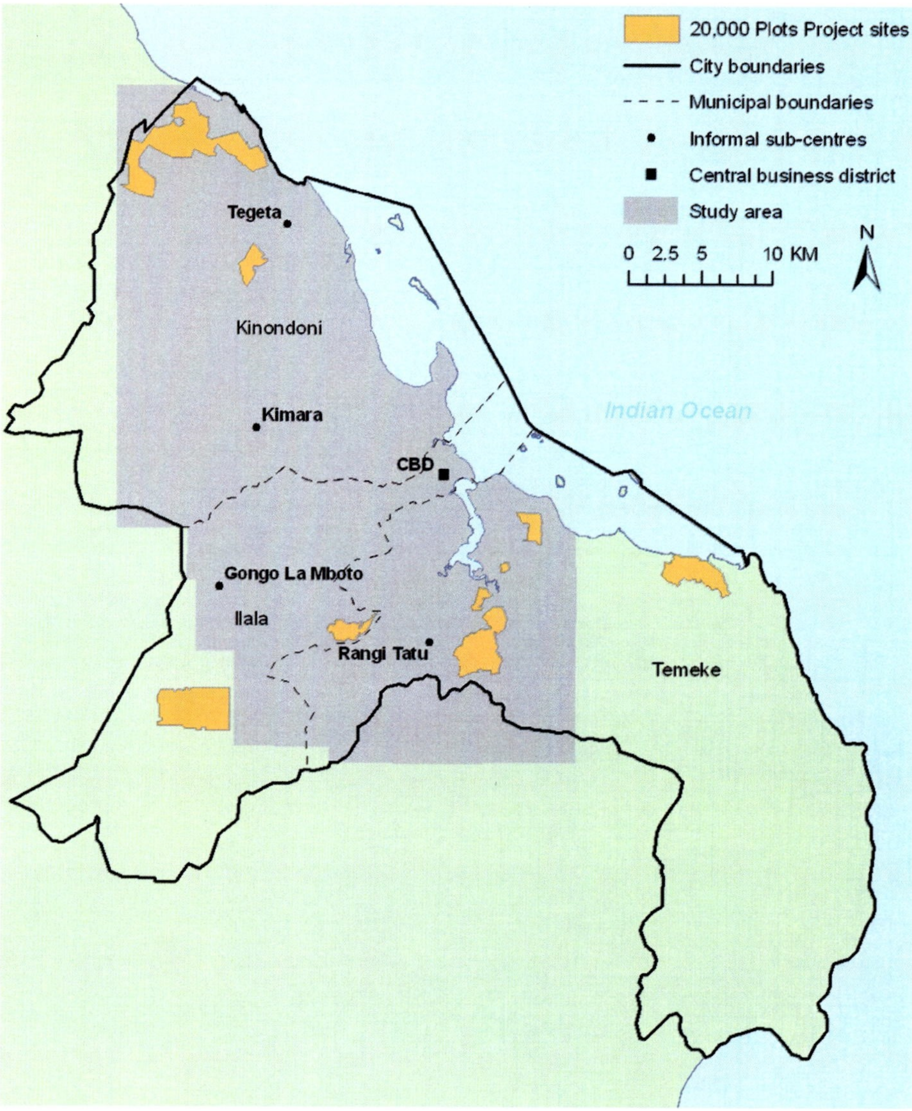

(Kombe & Kreibich 2006; A. Lupala 2002; J. Lupala 2002). Kironde (2006: 463) reports about calculations based on property tax databases which support that even more than 80% of all buildings in the city of Dar es Salaam are located in unplanned areas. Considering the higher house occupancy rates in unplanned areas, he argues that the proportion of the city's population living in unplanned areas is likely to be higher than 80%.

Urban growth in Dar es Salaam is mainly driven by informal residential location decisions by the urban poor and medium- and high-income groups. Research has brought forward that informal settling is basically driven by plot availability and plot affordability, thus, differing from formal

Figure 3: Built-up land development in Dar es Salaam for 1982, 1992 and 2002

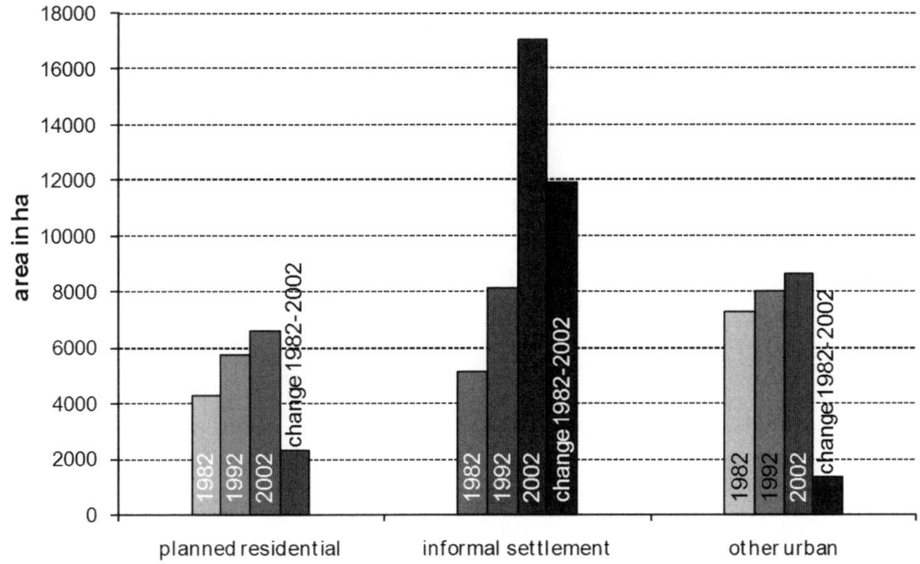

Source: Own illustration; calculations based on land-use data provided by ITC, Enschede; updated for 2002 by IRPUD, Dortmund

markets in developed countries (Kreibich et al. 2008; Lupala 2002). The imperfect market conditions are basically due to limited availability of information, ambiguous institutional settings and financial restrictions of informal settlers. Despite being informal, the location decision processes exhibit specific individual location rationalities. Kreibich even argues that the lack of statutory planning is to some degree compensated by "local institutions which join in social regulation trying to streamline the self interest of settlers in such a way that individual behaviour turns out to be beneficial for their community" (Kreibich 2010: 43). In the end, this invisible hand creates large parts of the spatial patterns observable on the city level.

According to empirical research in Dar es Salaam access to roads constitutes one of the most important drivers for allocation decisions of urban settlers (Hill et al. 2010). As a consequence the modelling work and database compilation focused on variables representing issues such as transport costs and distance to markets (reflecting location and agglomeration theories) and neighbourhood effects (in accordance with diffusion theory) to be relevant for informal location decisions and their urban dynamics.

Model development

The Dar es Salaam land-use model has been designed along the principles of CA and developed based on standard GIS software. Population projections by the UN (2008) have been used to determine the future demand for residential land. Variables identified as influencing urban growth have been integrated as major drivers to simulate future urban growth.

Database

Dar es Salaam like many other large cities in developing countries must be characterised as a data-poor environment. Public authorities lack up-to-date and accurate data and thus each public body basically works (literally) 'with their own homemade' reference datasets. Given the lack of personnel capacities and a missing tradition in database conception and maintenance even existing data created for instance in the course of international cooperation projects often gets lost after those projects end. Consequently, appropriate data for the modelling exercise was lacking. The compilation of the model database turned out to be quite demanding particularly concerning spatial coverage, multi-temporal availability, reliability and spatial resolution. Finally, initial datasets, e.g. land use and transport network, have been acquired from a partnering European research institution (ITC, Enschede, The Netherlands) and were subsequently updated and extended where possible based on aerial imagery. Based on these datasets a calibration database was compiled which included variables covering the following drivers: current land use, natural conditions, accessibility and neighbourhood relations. Neighbourhoods were defined with different radii around the core cell and calculated as simple counts as well as scores[1]. Table 1 shows the range of initial variables which have either been directly available or generated for the years 1982, 1992 and 2002 and which have subsequently been analysed and tested using binary logistic regression. This set of variables reflects data restrictions on the one hand but on the other hand underlines the aim to test a broader range of variables which all may have explanatory power for informal development processes. For example for accessibility a variety of datasets was generated to explore which variables would be able to best explain informal residential dynamics. Access to CBD with its concentration of formal functions is assumed to be an important but not the only location factor impacting on a cell's suitability. As a consequence, access to the transport network was added by means of distance to the next minor, major or any road. Moreover, visual analysis of interim simulation results and comparison with the real word land-use patterns exhibited that developments around informal sub-centres was underestimated by the model. Thus, accessibility of the four major informal sub-centres at the urban periphery was included (for the location of these informal sub-centres see Figure 2). Land prices could not be included in the model explicitly but many of the factors determining land prices in Dar es Salaam (A. Lupala 2002: 251f) have been incorporated.

Calibration and validation

The calibration procedure constitutes a central step in model development. The study area comprised about 97,000 cells (100 by 100 m) of which about 11,900 cells had transformed from vacant cells into informal residential land use between 1982 and 2002. This transformation constituted the dependent variable and in an iterative process using a binary logistic regression approach preselected sets of independent variables have been explored to explain

[1]Scores incorporate a distance-based weighting mechanism accounting for the fact that closer cells are assumed to have a larger impact on the core cell than more remote ones.

Table 1: Overview of variables included in the calibration database

Variable	Dynamic *	Description
Land use	yes	Land use of each cell (informal residential, planned residential, other urban, vacant / agriculture)
Elevation	no	Elevation above sea level of each cell
Slope	no	Slope in each cell in degrees derived from digital elevation model
Trunk road distance	yes	Euclidean distance from cell centre to next major road
Minor road distance	yes	Euclidean distance from cell centre to next minor road
All road distance	yes	Euclidean distance from cell centre to any next road
Network distance to CBD	yes	Shortest path distance in metres on transport network from each cell centre to the central business district
Travel time to CBD	yes	Average travel time (in minutes) on transport network from each cell centre to the central business district
Travel time to Tegeta	yes	Average travel time (in minutes) on transport network from each cell centre to the informal service centre of Tegeta
Travel time to Kimara	yes	Average travel time (in minutes) on transport network from each cell centre to the informal service centre of Kimara
Travel time to Gongo la Mboto	yes	Average travel time (in minutes) on transport network from each cell centre to the informal service centre of Gongo la Mboto
Travel time to Rangi Tatu	yes	Average travel time (in minutes) on transport network from each cell centre to the informal service centre of Rangi Tatu
Informal residential neighbourhood (count and score)	yes	Cellular neighbourhood considering informal residential land use: count and weighted count (score) of informal residential cells within circular neighbourhood around each cell
Planned residential neighbourhood (count and score)	yes	Cellular neighbourhood considering planned residential land use: count and weighted count (score) of formally planned residential cells within circular neighbourhood around each cell
Other urban neighbourhood (count and score)	yes	Cellular neighbourhood considering other urban land use: count and weighted count (score) of other urban cells within circular neighbourhood around each cell

* Multi-temporal data are marked to be dynamic when the specific dataset for the relevant years 1982, 1992 and 2002 changes between the time slices. In contrast, some datasets remain fixed as basically natural assets like slope and elevation do not change during the simulation period.

dynamics of the dependent variable. The model's prediction power was checked using a hybrid approach of quantitative degree of fit measures and visual interpretation of calibration results in comparison to real world development. The optimum model results were achieved with the set of independent variables as can be seen in Table 2.

This set of variables along with the coefficients is used by the model for calculating the transition potential of each cell. The calibration run based on this parameter set revealed a degree of fit with the real world development for the period of 1982 to 2002 of 61%. The results in terms of spatial allocation generated by the simulation model are displayed in Figure 4. They illustrate that the overall spatial growth patterns between 1982 and 2002 can be adequately reproduced by the model based on the selected parameter set. In order to validate the model subsequently to the calibration two validation runs have been conducted confirming the model's ability to adequately reproduce urban development for the tested time spans.

Model calibration revealed that accessibility constitutes a very important determinant for informal residential location decisions in Dar es Salaam. Although access to CBD is definitely relevant the analysis shows that rather a multi-facetted combination with travel time to informal sub-centres and the airline distance to next road explains the observed development patterns. In essence, it becomes evident that relevant factors interact in a complex and non-linear way. Neighbourhood effects introduce local-scale dynamics that seem to be interwoven with structural urban features and patterns both on the sub-centre and city-wide level.

Model design

This model has been developed and implemented using ArcGIS. It features a modular design being composed of several sub-models created either in ModelBuilder or Python. The core script drives the whole model and links together all sub-models. Data is fed into the model as GIS datasets. Output is also generated by the model in form of GIS data which can then be used directly in respective applications for visualisation and results assessment.

Table 2: Final variable set and coefficients

Variable	Coefficient
Intercept	3.793
Slope	-0.08963
Informal neighbourhood cells score (600 m radius)	0.006308
Planned neighbourhood cells score (600 m radius)	-0.004061
Travel time to CBD	-0.0555
Distance to next road	-0.0008854
Travel time to Gongo la Mboto sub-centre	-0.02183
Travel time to Rangi Tatu sub-centre	-0.01203
Travel time to Tegeta sub-centre	-0.01980
Travel time to Kimara sub-centre	-0.002483

Figure 4: Land-use simulation result for 2002 with the final calibrated parameter set

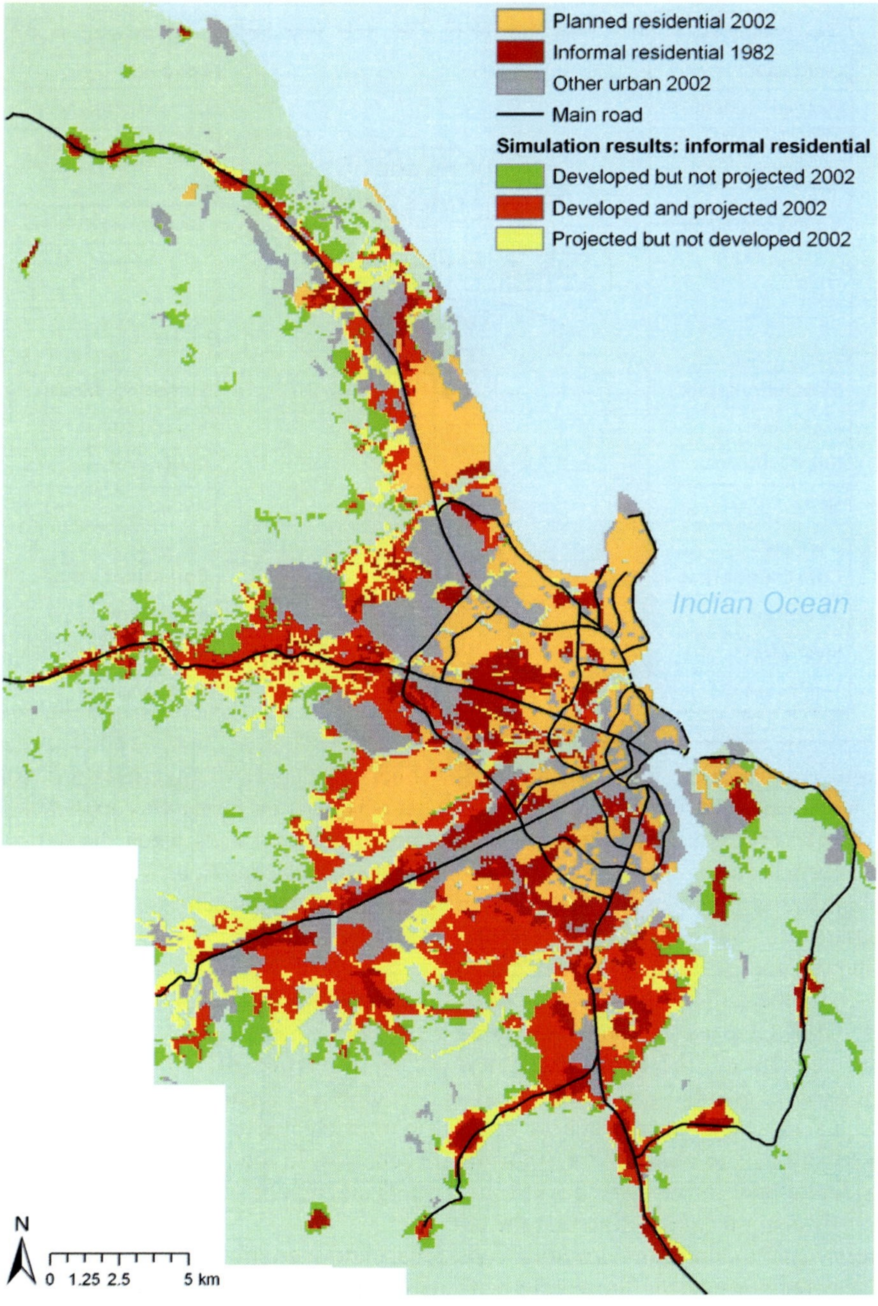

At the beginning of each model run the vacant cells being subject to the modelled transformation process of informal urbanisation are determined (see Figure 5). For each one of these cells suitabilities are calculated based on the factors identified during model conception and calibration. Further-

Figure 5: Land-use transformation process

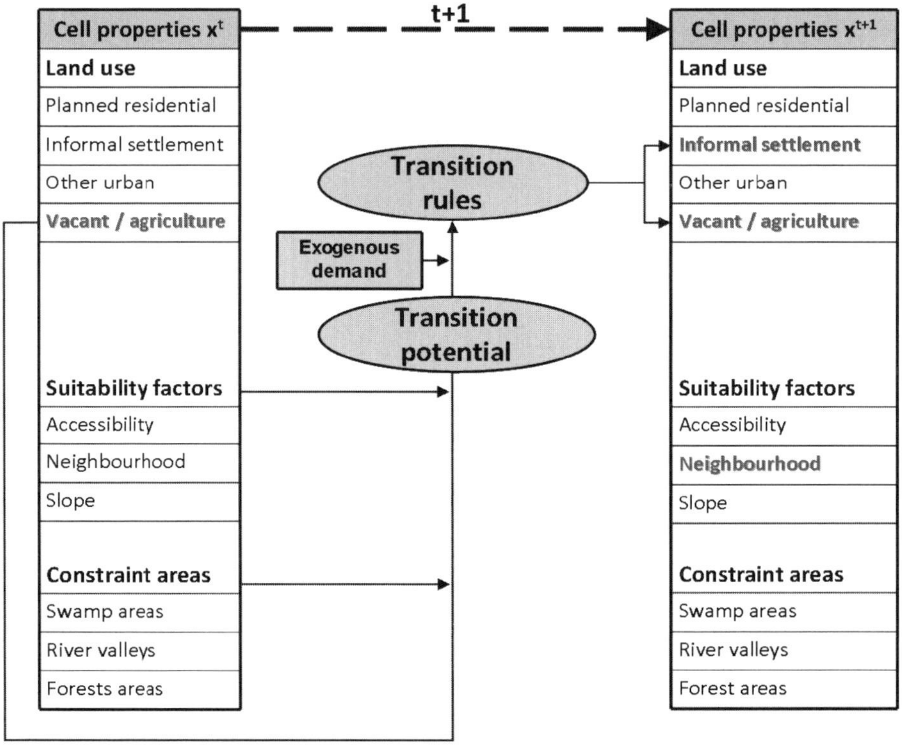

more, cells located in constraint areas, i.e. swamp areas, river valleys and forests reserves, are excluded from the transformation process since no development is assumed to take place in these areas. Subsequently, the transition potential for each cell is calculated based on the suitability factors. Considering the land-use demand computed exogenously, i.e. the demand for new informal residential cells, the model assigns as many vacant cells as needed to satisfy the demand for informal residential use in the subsequent timestep. Transition rules define that the respective number of transformation cells is selected considering those cells with the highest transition potential. Based on the assumption that the model developed is already subject to a multitude of abstractions and assumptions implying a certain degree of fuzziness no explicit perturbation factor was applied. The model iteration ends when the demand for cell transformations is satisfied. The remaining vacant cells are not subject to transformation.

Scenario application

The simulation of future development perspectives includes a baseline scenario and selected transport infrastructure scenarios. In the following scenario details and results will be presented.

Baseline scenario

To demonstrate likely perspectives of future urban growth the simulation model was applied to forecast informal urban expansion. The baseline scenario extrapolates past trends into the future,

Figure 6: Baseline scenario results for 2012 and 2022

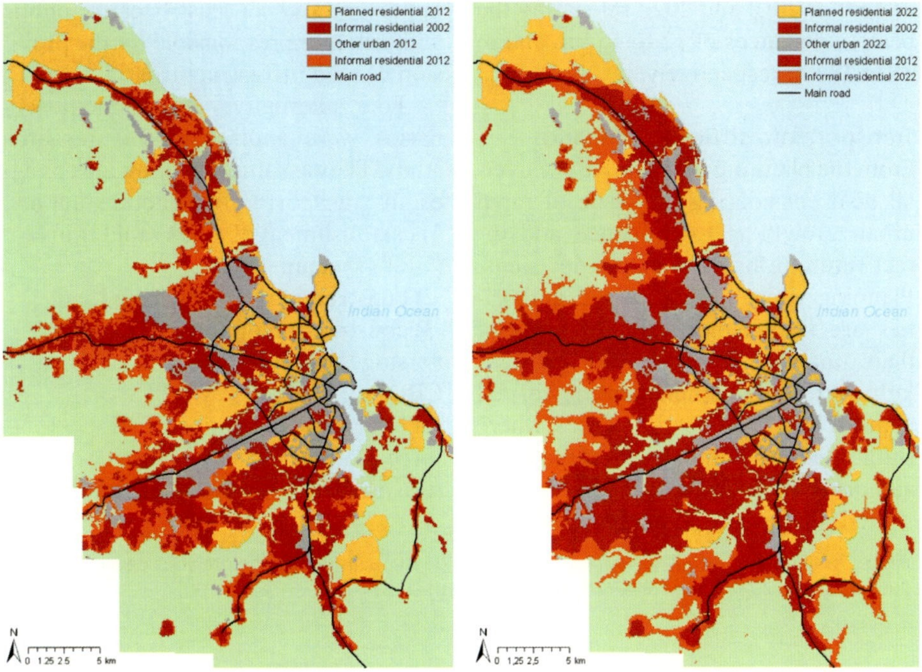

its simulation results can be seen in Figure 6 showing the land-use distribution for the years 2012 and 2022. The urban expansion is mapped as projected to take place until 2022 implied by adding 21,322 ha of informal residential use and designated 20,000 Plots Project sites to the urban fabric of the study area. The demand for newly developed informal residential land was derived from UN population projections (UN 2008) and model assumptions made on population densities in informal and planned settlements and zoning designations for the latter land-use type by the 20,000 Plots Project.

The baseline scenario results clearly illustrate the spatial dimension and distribution of settlement pressure which Dar es Salaam is likely to face in the near future. The basic development trends can briefly be summed up for the decade between 2002 and 2012 as a continuous urban expansion of the ribbons alongside the main arterial roads towards the urban periphery. The subsequent decade is projected to experience that settlements will encroach the so far less developed hinterlands between the arterial roads filling up the interstitial areas. The forecasts of the baseline scenario confirm the assumption that Dar es Salaam will continue to rapidly expand its urban borders. Figure 7 demonstrates the general trend of acceleration of urban expansion and the temporal shift of peaks of informal urban dynamics according to concentric rings around the CBD. In the decade between 1982 and 1992 the largest dynamics could be observed at distances between 6 and 9 km from the CBD. During the next decade locations at this distance also experienced rapid development but the peak of transformations shifted to distances between

9 and 12 km. A similar effect is forecasted for 2012 and 2022 relocating the peak to distances of 12 to 15 km and to 15 to 18 km respectively.

Transport infrastructure scenarios

From the planning perspective the overall goal considering the issue of rapid urban growth is to influence and direct future informal location decisions. Planning capacities of public authorities are limited not only in terms of their financial resources but particularly in terms of implementing formal plans. Thus, the question arises where suitable access points for local planning authorities exist. Research has demonstrated that one approach to provide more favourable conditions to informal urban settlers and, thus, guide future urban growth is to provide trunk infrastructure (Hill et al. 2010) - particularly road infrastructure (Hill & Lindner 2010) - as accessibility is one main key to informal urbanisation. Consequently, transport infrastructure scenarios constitute an adequate approach to demonstrate consequences of planning interventions and alternative development opportunities (not least because public authorities are responsible for the provision of road infrastructure).

Four alternative development scenarios were applied in this research study. They assume the following changes in transport infrastructure supply:
1) establishment of a bus rapid transit (BRT) system;
2) construction of a new ring road;
3) construction of a new bridge traversing the creek south-east of the CBD;
4) joint implementation of all three aforementioned infrastructure projects (combined scenario).

Figure 7: Number of newly developed informal residential cells during the previous decade for 1992, 2002, 2012 and 2022 along concentric rings around the CBD

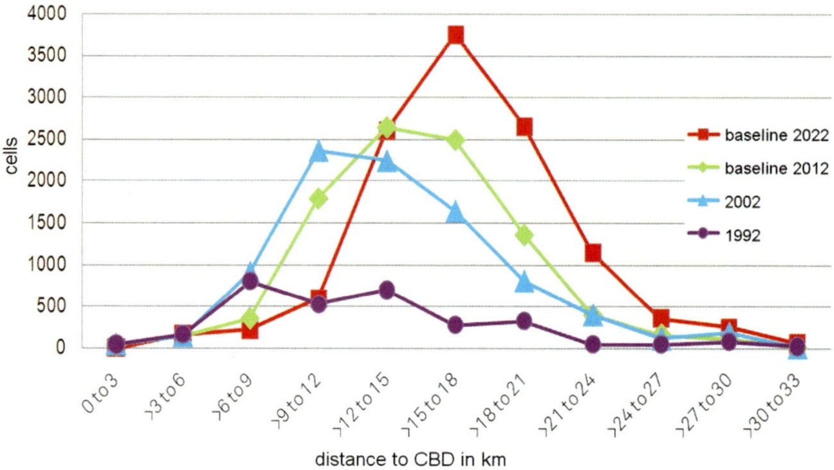

All these measures are part of the current Dar es Salaam Transport Masterplan. The masterplan aims at formulating short-term actions to alleviate current traffic problems and provide a long-term perspective for urban transport policy and development in the Dar es Salaam metropolitan area until 2030. Most of the following specifications concerning infrastructure rojects are based on the draft final report for this plan (DCC 2008).

The combined scenario brings together the integrated establishment of the three stand-alone transport projects. The routing of additional road sections and modified classification of specific sections in terms of improved travel speed induced by the BRT system can be seen in Figure 8. The simulation run forecasts the concomitant land-use distribution for 2022 as illustrated in the map on the left-hand side. On the right hand side the changes in land-use transformation compared to the baseline scenario are displayed.

The combined scenario reveals a clear trend of redistributing development dynamics towards the areas south-east of the creek relieving urban pressure from large parts of the city's peri-urban areas. Accessibility improvements for the area south-east of the creek will be realised basically through the construction of a bridge and a new ring road which will mutually amplify the positive effect. In combination with those settlements planned formally that area is projected to be almost completely transformed to urban uses.

In addition to the cell-based map comparison Figure 9 illustrates the impacts of transport projects according to airline distance to the CBD. The BRT scenario causes redistribution of newly developing cells to more remote locations with a peak of development at 21 to 24 km. The ring road scenario redistributes development from areas of 15 km on outwards to nearer distances with a peak at 6 to 9 km. The bridge scenario and also the combined scenario show similar pattern strongly

Figure 8: Combined transport scenario results for 2022 (left) and land-use changes compared to baseline scenario (right)

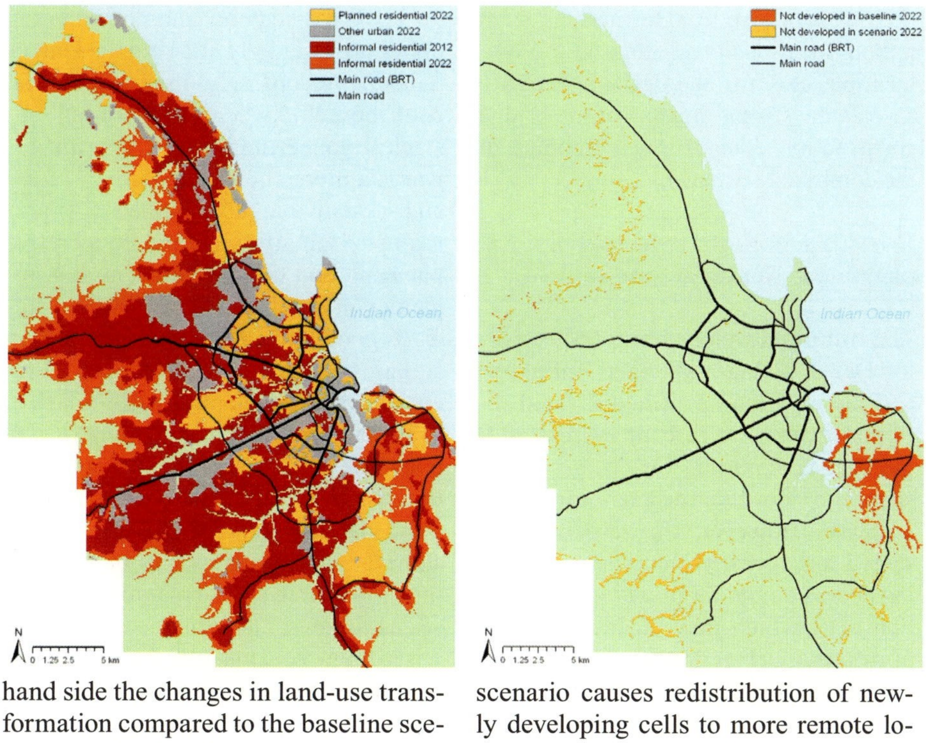

Figure 9: Distribution differences between the combined scenario and the baseline scenario results in 2022 along concentric rings around the CBD

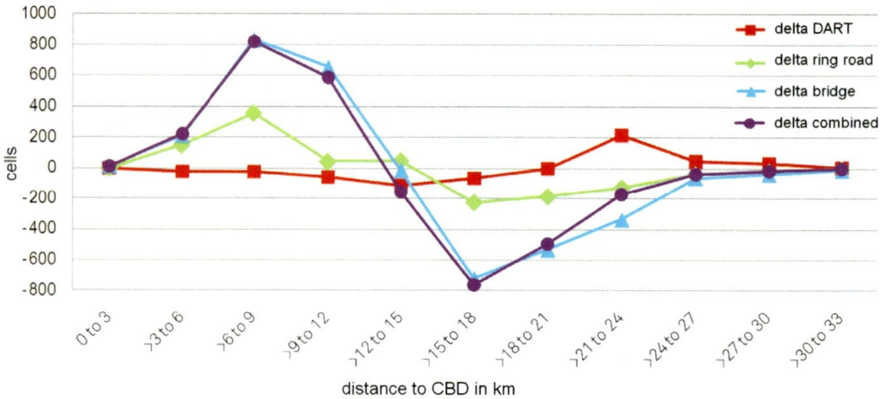

shifting land-use transformations from remote locations at 12 km and more to locations closer to the CBD. This figure reveals the strong impacts of the bridge construction even in the outcomes of the combined scenario.

Towards a more strategic spatial planning in Dar es Salaam

The future development perspectives for Dar es Salaam sketched out above and the projected continuous rapid informal expansion call for more strategic spatial planning. So far, a cooperative and coordinating planning is not in place. However, the research presented has revealed that some promising access points for planning intervention exist which may enable guidance of informal urban growth. Urban models like the simulation model presented may become a suitable tool or platform to overcome some of the existing coordination gaps and to strengthen strategic planning.

In general, spatial models are increasingly employed as decision support tools in urban planning in order to inform planners and decision makers. So far cross-sectoral and coordinating planning is largely missing in Dar es Salaam. Accordingly, the model could contribute to overcome planning obstacles pioneering the way forward towards a more coordinated urban future and sustainable development. A prerequisite for informed and cooperative planning and the development and implementation of a sustainable urban policy is good knowledge of the status quo situation and the major trends and drivers of urban development. Here, urban information and decision support systems come into play. Such tools serve to identify recent urban development trends and drivers of change. Lowry summed up the advantages of integrating models into spatial planning processes as follows: "Not that computers are wiser than their masters, but rather (...) they perform the most monotonous and repetitive temporal tasks at high speed and with absolute mechanical accuracy. The model-builder can make use of this capacity only insofar as he is able to perceive repetitive temporal patterns in the processes of urban life, fixed spatial relationships in the kaleidoscope of urban form" (Lowry 1965: 158 f).

Moreover, simulation models serve to forecast future states of urban systems illustrating the needs for future planning and investment. They can be used to demonstrate the possible impacts of different planning measures and, thus, support participatory approaches in urban planning. As the "future orientation of planning is unique to the field's identity (...) the very substance of urban planning is founded in time" (Myers & Kitsuse 2000: 225) so that urban planning and urban models have to be discussed jointly.

Couclelis has argued that connecting these two spheres could "amplify the positive synergies between the two domains and enhance the ability of spatial planning to prepare for the future" (2005: 1353). Strategies are judged to be "the most fundamental aspect of plans (...) because strategies directly account for actions, outcomes, intentions, and uncertainty. Strategies address most completely the difficulties created by interdependence, indivisibility, irreversibility, and imperfect foresight" (Hopkins 2001: 6). Thus, the sound claim for more strategic planning has been revitalised during recent years by various scientists (e.g., Albrechts 2004; Friedman 2004; Salet & Faludi 2000). Albrechts argues that throughout the manifold efforts underway in Europe towards a more strategic planning for cities, city-regions and regions the motivations are varied but "the objectives have typically been to articulate a more coherent spatial logic for land-use regulation, for resource protection, for action orientation, for a more open type of governance, for introducing sustainability, and for investments in regeneration and infrastructure" (2004: 749). Nevertheless, planning and modelling practices exhibit a clear separation as "earlier hopes of creating a stronger, more scientific urban land-use planning practice through integration with models did not materialize" (Couclelis 2005: 1355). In this context Couclelis (2005) demands that appropriate urban models have to be developed that are suitable to step into the place when it comes to support this strategic and coordinating function of planning. Tools such as the presented land-use simulation model could be of particular importance for developing countries where the question of how to spend the scarce resources is even more demanding and should be addressed strongly strategically and in a coordinating manner.

Conclusions

The modelling exercise has identified key drivers for informal residential growth in Dar es Salaam. Accessibility factors have been identified to be important determinants. The CBD has proved to be still relevant for informal settlers but in a multi-facetted combination with travel time to informal sub-centres and the airline distance to next road to facilitate access to the road network at all. The neighbourhood factor has proven to replicate important drivers of informal land-use dynamics which cannot be captured explicitly. Whereas the proximity to existing informal settlements is largely attractive, regression analysis suggests that planned settlements rather tend to repulse informal development. The neighbourhood factor, therefore, seems to be highly suitable to compensate for data gaps and to account for small-scale dynamics that influence the informal land and housing markets and drive urban expansion to take place in a highly diffusive manner.

The above determinants seem to interact in a complex and non-linear way which can be successfully modelled using the CA approach. The neighbourhood effects introduce local-scale dynamics that represent bottom-up urban development processes. Nevertheless, they seem to be interwoven with structural urban features and patterns both on the sub-centre and city-wide level like accessibility of informal subcentres and the CBD. The CA-based model has succeeded in reproducing these relatively complex socio-spatial dynamics with a set of simple rules and limited data.

Out of the list of variables accessibility and transport infrastructure make up the key access point for planning intervention. Transport projects clearly reveal strong impacts on informal urban development and should be perceived as a guiding instrument and one strategic point for investment particularly considering the scarce resources available to planning authorities in Dar es Salaam. Moreover, knowledge about the impacts of transport infrastructure investments could break ground for a cross-sectoral and cooperative action of relevant stakeholders responsible for the provision of additional basic services and infrastructures like water and electricity supply, sewerage, health and education facilities, etc.

In addition to the aforementioned achievements some shortcomings resulting in need for further research can be formulated. The limited database puts some restrictions to the modelling work. The simulation results for the year 2022 demonstrated that the urban expansion of the city already reached the boundaries of the study area of this research which caused some redistributing effects. Thus, at least for simulations beyond this point in time an extended study area is required which of course impacts on the database needed. Figures on population distribution could largely enhance the models capability to better reflect the phenomenon under study. During their development informal settlements undergo significant consolidation processes when land is further subdivided as the demand for plots rises. These processes resulting in increased densification could not be modelled so far due to the lack of information on population densities. Significant improvements to the model can be expected by enabling it to cover and simulate these processes directly. Furthermore, the rather static transport network puts some obstacles to the ability of the model to reproduce urban dynamics realistically. An integrated transport model could substantially improve the simulation results. So far, the model is restricted to reproduce the transformation from vacant land to informal residential use only which is the predominant transformation observed. Including further land-use dynamics may provide additional insights into urban development mechanisms and planning intervention opportunities. Last but not least further research is needed on how to integrate such expert systems into planning practice with the aim to employ them as practical decision support systems. This would require establishing adequate structures in organisational and institutional as well as in technical terms to allow for a useful and effective integration and maintenance of the model. For sound analyses and interpretation of the model outputs capacity building is one fundamental prerequisite. Involving public, private and academic stakeholders into a modelled planning discourse could further enhance the integration of academic science and planning practice and eventually support rational plan-making at city level.

References

Albrechts, L. (2004): Strategic (spatial) planning reexamined. In: Environment and Planning B: Planning and Design, 31 (5), pp. 743-758.

Barredo, J., Demicheli, L., Lavalle, C., Kasanko, M. & N. McCormick (2004): Modelling future urban scenarios in developing countries: An application case study in Lagos, Nigeria. In: Environment and Planning B: Planning and Design, 31 (1), pp. 65-84.

Barredo, J., Kasanko, M., McCormick, N. & C. Lavalle (2003): Modelling dynamic spatial processes: Simulation of urban future scenarios through cellular automata. In: Landscape and Urban Planning, 64, pp. 145-160.

Batty, M. (1997): Cellular automata and urban form: a primer. In: Journal of the American Planning Association, 63, pp. 264-274.

Batty, M. & Y. Xie (1994): From cells to cities. In: Environment and Planning B: Planning and Design, 21, pp. 31-48.

Benenson, I. & P. M. Torrens (2005): A minimal prototype for integrating GIS and geographic automata systems. In: Atkinson, P. M., Foody, G. M., Darby, S. E. & F. Wu (eds.): GeoDynamics. CRC Press, Boca Raton, London, New York, pp. 347-367.

Bronger, D. (2004): Metropolen, Megastädte, Global Cities. Die Metropolisierung der Erde. Wiss. Buchgesellschaft, Darmstadt

Couclelis, H. (1988): Of mice and men: What rodent populations can teach us about complex spatial dynamics. In: Environment and Planning A, 20(1), pp. 99-109.

Couclelis, H. (1997): From cellular automata to urban models: New principles for model development and implementation. In: Environment and Planning B: Planning and Design, 24(2), pp. 165-174.

Couclelis, H. (2005): Where has the future gone? Rethinking the role of integrated land-use models in spatial planning. In: Environment and Planning A, 37(8), pp. 1353-1371.

DCC (Dar es Salaam City Council) 2008: Transport Policy and System Development Master Plan - Draft final report. Dar es Salaam.

de Soto, H. (2000): The mystery of capital: Why capitalism triumphs in the west and fails everywhere else. Basic Books, New York.

Friedmann, J. (2004): Strategic planning and the longer range. In: Planning Theory and Practice, 5(1), 49-67.

Fuchs, R. J., Brennan, E., Chamie, J., Lo, F.-C. & J. I. Uitto (eds.) (1994): Mega-city growth and the future. United Nations University Press, Tokyo.

Hall, P. & U. Pfeiffer (2000): Urban future 21: A global agenda for 21st century cities. E & FN Spon, London.

Hill, A., Hühner, T., Kreibich, V. & C. Lindner (2010): Dar es Salaam, megacity of tomorrow - Informal urban expansion and the provision of technical infrastructure. In: Kraas, F. (ed.): Megacities - Our global urban future. Springer, Series: International year of planet Earth (forthcoming).

Hill, A. & C. Lindner (2009): City structure and development of Dar es Salaam, Tanzania – Patterns and trends of informal urban dynamics. In: Hill, A. & K. Weiß (eds.): Spatial Planning with Obstacles. 40 Years of Spatial Planning in Crisis Regions and Developing Countries. Dortmunder Beiträge zur Raumplanung 133, Dortmund: Institut für Raumplanung, pp. 55-67.

Hill, A. & C. Lindner (2010): Modelling informal urban growth under rapid urbanisation - A CA-based land-use simulation model for the city of Dar es Salaam, Tanzania. Doctoral thesis, TU Dortmund University.

Hopkins L. D. (2001): Urban development: The logic of making plans. Island Press, Washington, DC

Kironde, J. M. L. (2006): The regulatory framework, unplanned development and urban poverty: Findings from Dar es Salaam, Tanzania. In: Land Use Policy, 23, pp. 460–472.

Kombe, W. J. (2009): Women access to land and property in sub-Saharan Africa: Policies, practices and the role of the informal land supply sector. In: Hill, A. & K. Weiß (eds.): Spatial planning with obstacles. 40 years of spatial planning in crisis regions and developing countries. Dortmunder Beiträge zur Raumplanung 133, Dortmund: Institut für Raumplanung, pp. 31-54.

Kombe, W. & V. Kreibich (2006): The governance of informal urbanisation in Tanzania. Mkuki na Nyota Publishers, Dar es Salaam.

Kraas, F. & G. Mertins (2008): Megastädte in Entwicklungsländern – Vulnerabilität, Informalität, Regier- und Steuerbarkeit. Geographische Rundschau, 11/2008, pp. 4-10.

Kreibich, V. (2010): The mode of informal urbanisation – Reconciling social and statutory regulation in urban land management. In: McFarlane, C. & M. Waibel (eds.): Urban informality in a globalising world: reflections on the formal and informal. Hoboken, Oxford (forthcoming).

Kreibich, V., Hill, A., Hühner, T. & C. Lindner (2008): Informelle Stadterweiterung und die Versorgung mit technischer Infrastruktur in der zukünftigen Megacity Dar es Salaam, Tansania. Geographische Rundschau, 11/2008, pp. 12-19.

Lowry, I. S. (1965): A short course in model design. In: Journal of the American Planning Association, 31(2), 158-166.

Lupala, A. (2002): Peri-urban land management for rapid urbanisation. The case of Dar es Salaam. SPRING Research Series 32, Dortmund.

Lupala, J. (2002): Urban types in rapidly urbanizing cities - Analysis of formal and informal settlements in Dar es Salaam, Tanzania. Doctoral Thesis submitted to the Department of Infrastructure, Royal Institute of Technology, Stockholm.

May, R. M. (1976): Simple mathematical models with very complicated dynamics. In: Nature, 261, 459-467.

Ménard, A. & D. J. Marceau (2005): Exploration of spatial scale sensitivity in geographic cellular automata. In: Environment and Planning B: Planning and Design, 32 (5), 693-714.

Mertins, G. (1992): Urbanisierung, Metropolisierung und Megastädte. Ursachen der Stadt"explosion" in der Dritten Welt. In: Deutsche Gesellschaft für die Vereinten Nationen (Ed.): Mega-Städte – Zeitbomben mit globalen Folgen. Dokumentationen, Informationen, Meinungen, 44, Bonn, pp. 7-21.

Myers, D. & A. Kitsuse (2000): Constructing the future in planning: A survey of theories and tools. In: Journal of Planning Education and Research, 19, pp. 221-231.

Roy, A. (2005): Urban informality – Towards an epistemology of planning. In: Journal of the American Planning Association, 71 (2), pp. 147-158.

Roy, A. & N. Al Sayyad (2004): Prologue/Dialogue. Urban informality: Crossing borders. In: Roy, A. & N. Al Sayyad (eds.): Urban informality: Transnational perspectives from the Middle East, Latin America, and South Asia. Lanham, pp. 1-6.

Salet, W. & A. Faludi (2000): Three approaches to strategic planning. In: Salet, W. & A. Faludi (eds.): The revival of strategic spatial planning. Royal Netherlands Academy of Arts and Sciences, Amsterdam, pp. 1-10.

Schelling, T. C. (2006): Micromotives and macrobehavior. Norton & Company, New York.

Sietchiping, R. (2004): A Geographic Information System and Cellular Automata-Based Model of Informal Settlement Growth. Doctoral thesis, The University of Melbourne.

Torrens, P. M. (2000): How cellular models of urban systems work (1. Theory). CASA Working Paper 28, Centre for Advanced Spatial Analysis, University College London, London.

Torrens, P. M. & D. O'Sullivan (2001): Editorial: Cellular automata and urban simulation: Where do we go from here? In: Environment and Planning B: Planning and Design, 28 (2), pp. 163-168.

UN (United Nations) 2007: World population prospects: The 2006 revision. New York.

UN (United Nations) 2008: World urbanization prospects: The 2007 revision - Highlights. New York.

UN (United Nations) 2010: World Population Prospects: The 2008 Revision and World Urbanization Prospects: The 2009 Revision, available at http://esa.un.org/wup2009/unup/, (9.6.2010).

UN-Habitat (United Nations Human Settlements Programme) 2009: Global report on human settlements 2009: Planning sustainable cities. Earthscan, Sterling, Nairobi.

URP (Department of Urban and Regional Planning) 2006: The implication of twenty thousand plots project on the emerging city form of Dar es Salaam. Unpublished report, Dar es Salaam.

von Bertalanffy, L. (1968): General system theory – Foundation, development, application. George Braziller, New York.

Wegener, M. (2004): Overview of land-use transport models. In: Hensher, D. A. & K. Button (eds.): Transport Geography and Spatial Systems. Handbook 5 of the Handbook in Transport. Pergamon/ Elsevier Science, Kidlington, pp. 127-146.

White, R. & G. Engelen (1993): Cellular automata and fractal urban form: A cellular modelling approach to the evolution of urban land-use patterns. In: Environment and Planning A, 25, pp. 1175-1199.

White, R. & G. Engelen (1994): Cellular dynamics and GIS: modelling spatial complexity. In: Geographical Systems, 1, pp. 237-253.

White, R., Engelen, G., Uljee, I., Lavalle, C. & D. Erlich (1999): Developing an urban land use simulator for European cities. In: Fullerton, K. (ed.): Proceedings of the 5th EC GIS workshop: GIS of tomorrow. European Commission, Joint Research Centre, pp. 179-190.

Wolfram, S. (1984a): Preface. Physica, 10D, pp. vii–xii.

Wolfram, S. (1984b): Universality and complexity in cellular automata. Physica, 10D, pp. 1-35.

Wolfram, S. (1988): Complex systems theory. In: Emerging syntheses in science: Proceedings of the Founding Workshops of the Santa Fe Institute, Addison–Wesley, Reading, MA., pp. 183–189.

Simulation of urban land use change in North Rhine-Westphalia (Germany) with the Java-based modelling platform Xulu

Roland Goetzke und Michael Judex

Introduction

Land use change and land use modelling

A main aspect of global environmental change is the alteration of natural land cover by human use (Vitousek et al. 1997). In Central Europe more or less every parcel of land is in some way used by humans, e.g. for settlement, for farming, or for recreation. Especially in the densely populated regions like the German federal state of North Rhine-Westphalia (528 inh. / km²), land is a limited resource. During the last decades cities grew constantly. The space that is used per person has doubled in the last fifty years, while in the same time the available agricultural area has decreased. More and more this fact is seen as a problem, so that the conservation of the European landscape has reached the political agenda by now (European Environment Agency 2007; Antrop 2004). Many land use decisions that are made today will have an effect for many decades in the future. Thus, it is important to provide tools to decision makers that enable them to simulate possible developments of land use changes as consequences of their decisions. Land use models are suitable tools: On the one hand, they generate a deeper knowledge of the land use system and the interactions between land use and different driving forces, and on the other hand they can provide decision makers a basis for discussions by calculating different scenarios of land use change (Verburg et al. 2004).

Problems of model selection and model comparison

The quantity of existing land use models existing opens up a variety of options for the model user to choose between different methods and theoretical concepts that best fit to the real conditions in the study area. But this quantity is both a blessing and a curse. Oftentimes, certain aspects of a land use model perfectly represent specific structures or processes in the real world, while others are hardly considered. A model remains a simplified and generalised representation of reality (Mulligan & Wainwright 2004), so that one has to be satisfied with concentrating on the most important parts of the land use system (Lambin 2004). The intrinsic nature of simplification in land use models leads to the fact that certain aspects, which may be of major interest for the researcher, cannot be represented by an existing model, which then leads in turn to the creation of a variety of modelling approaches. That makes it difficult even to classify models in main groups, although efforts of classifying modelling approaches were conducted by i. a. Briassoulis (2000) and Verburg et al. (2003, 2004). Verburg et al. (2006) compare and contrast certain model properties, which results in a very detailed characterisation of modelling approaches. The intended modelling purpose and the available data normally define the choice of a certain model. In general, the modeller wants the model to behave most similar to the real world system, which is why the model of choice is usually the one with the highest predictability (Pontius Jr & Malanson 2005). According to Pontius et al. (2004) a model should always be evaluated by comparing its predicted results in quantity and location with a Null-Model. A Null-Model is a model that predicts no change. If, for example, 10 % of the land use in a map changes between time t1 and t2 into another category, a Null-Model would have a predictability of 90 %, because 90 % of the land use does not change. Thus, different models can be compared regarding their validity in relation to a Null-Model. The most extensive model comparison so far has been conducted by Pontius et al. (2008) in which 9 established land use models have been tested. By a comparative evaluation of models it is certainly up to the modeller to interpret, if the performance of a model is related to the modelling algorithm, to the data availability, or to the assumptions that have been formulated about the reality. Because of the diverging complexity of different modelling approaches, potential errors in the modelling process may occur due to completely different reasons.

In this study we compare two modelling approaches that are based on different theoretical frameworks against the background of the simulation of urbanisation processes. The models are intended to simulate urban growth (or change) processes in the heterogenous settlement pattern of a large, highly urbanised study area.

Methods

Study area and Data

Study area is the German federal state North Rhine-Westphalia (NRW), an area of about 34,000 km², which is quite large for the application of a regional land use model. NRW contains 396 municipalities and is the state with the highest population density in Germany. 20 % of the total area can be considered as built-up land. Especially the Ruhr area and the Cologne-Bonn region are widely urbanised. About 50 % of the land is under agricultural land use and a quarter of the total area is forested. This state shows a broad range of land use changes. The Ruhr area, an old industrial area shaped by the coal and steel industry in the state's centre experiences large structural changes since the 1960's, while along the Rhine river a number of urban growth centres are located. In the rural regions, especially close to the metropolitan areas, a high gain in settlement areas can be denoted. Figure 1 shows the population development and thus gives evidence about the regions with increase or decrease of population.

In order to visualise the problem of the increase of impervious surfaces in NRW four land use classifications based on Landsat satellite data have been produced on behalf of the Ministry of Environment, Nature Conservation, Agriculture and Consumer Protection of NRW, covering the years 1975, 1984, 2001 and 2005. The classification procedure is reported at Schöttker et al. (2004) and Goetzke et al (2006). Figure 2 shows the classification results for 2005 and highlights the heterogeneous structures of land use in NRW. This data set was available at a spatial resolution of 30 m x 30 m and contained 12

Figure 1: Map of the population development in NRW between 1984 and 2001

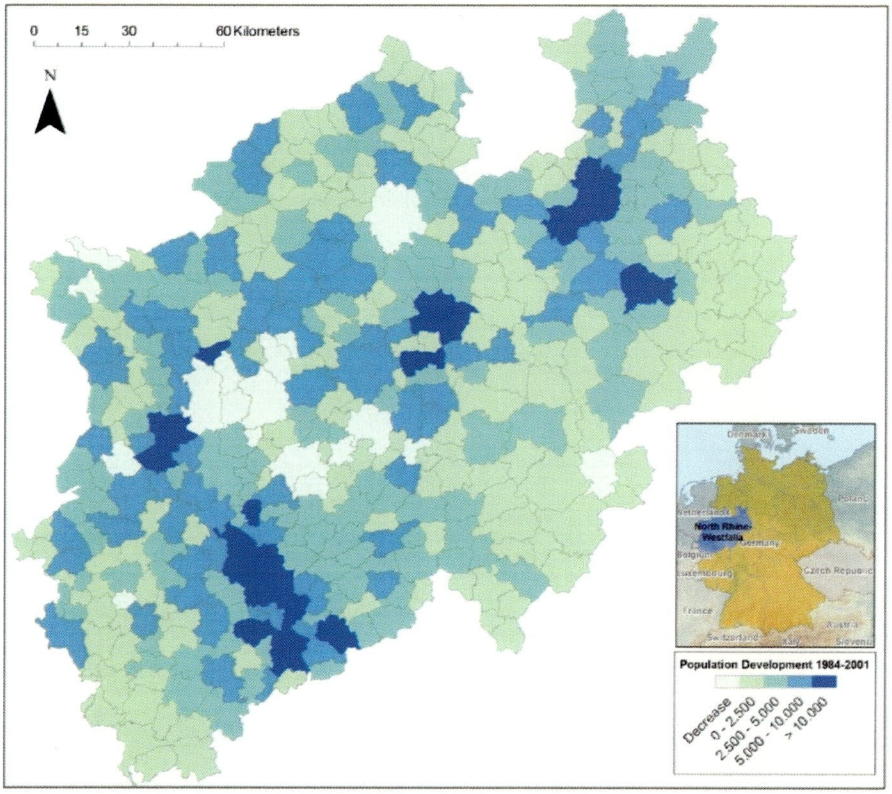

land use classes. For the modelling approach presented in this work, the resolution had to be reduced to 100 m due to computational power, which in fact reduced the amount of pixels that had to be computed by the modelling algorithm from 38 to 3.4 Millions. Moreover, as we will show later, land use models are rarely able to predict changes accurately on such a high resolution. Additionally the land use classes have been reduced from 12 to 6 for the modelling approach, whereas three urban classes showing the degree of imperviousness have been aggregated to "built-up", three forest classes have been aggregated to "forest", and classes that feature special human use (lignite mining areas, gravel pit & quarries, military training areas) have been aggregated to "other". The aggregation of the different classes of imperviousness to a single built-up class seemed reasonable, because first of all our study concentrates on general patterns of urban land use change. Moreover, the urban growth model used in this study (see section below: Urban growth model based on Cellular Automata) is only able to predict the change of one single category, which is urban. The land use changes of the 6 aggregated land use categories between 1975 and 2005 in NRW are depicted in Figure 3.

For the time frame visualised in Figure 3, the built-up (urban) areas have increased from 8.7 % up to 14.2 % of the total land area and forest from 22.8 %

Figure 2: Land use classification for NRW for the year 2005

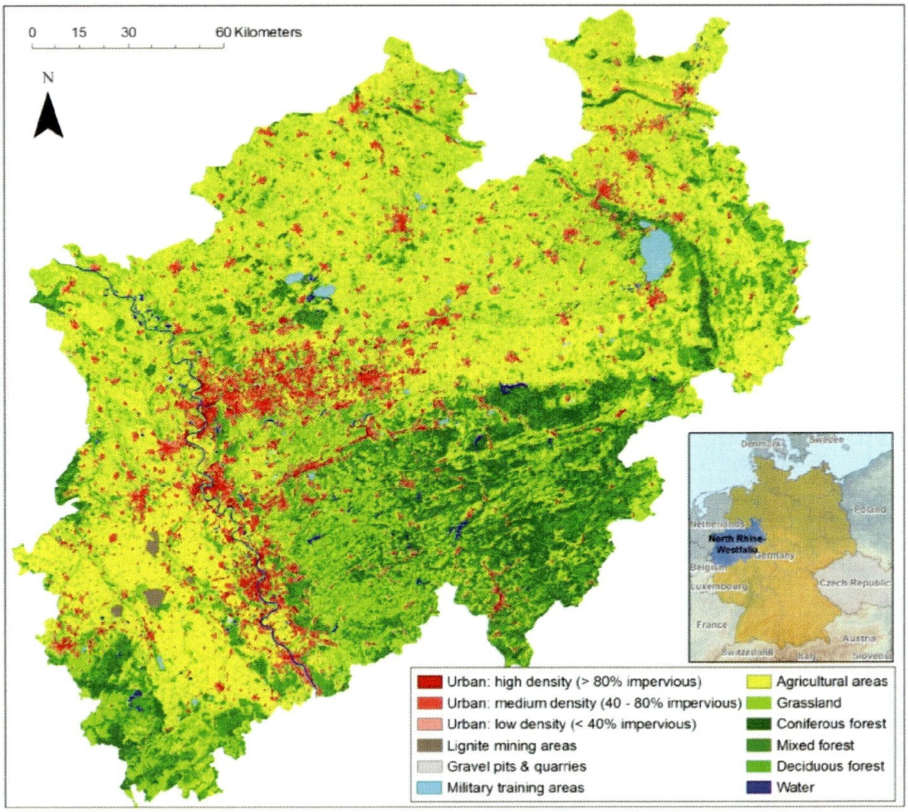

to 25.6 %, while agricultural areas have decreased from 38.1 % to 34.6 % and grassland from 28.8 % to 23.5 %.

To calibrate the land use models we used the classification results of the years 1984 and 2001. The data set of 1975 could not be included in this study because of the lack of availability of reliable socio-economic data for this time, which are needed for one of the modelling approaches. For the subsequent validation the map of 2005 has been used. Between 1984 and 2001 a net change os 2.8 % from non-built-up to built-up area occurred. In the short time span between 2001 and 2005 the same category changed by only 0.1 %.

Models used

Two different land use models are used to model land use change of the urbanised regions of NRW which follow different theoretical approaches. The first model is CLUE-s, a statistical model following an integrative system-oriented approach. That means the model considers the whole land system and relies on the competition between land use categories. The model is provided with the information of the approximate quantity of change for each land use category for every time step (in our case: 1 time step = 1 year), and the model calculates, which locations are most appropriate for the change of a certain land use category based on statistics.

Figure 3: Land use distribution in NRW 1975, 1984, 2001 and 2005

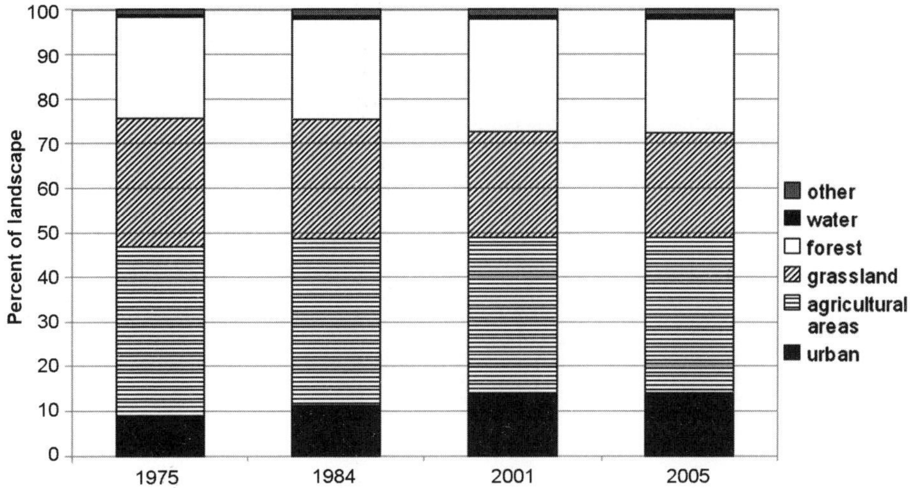

Hence, this model can be called an allocation model. The second model we used is a Cellular Automaton (CA) model, which simulates urban growth. Its functionality is very close to the SLEUTH urban growth model, which will be explained later.

Integrated system-oriented model (CLUE-s)

The CLUE-s model (Conversion of Land Use and its Effects at Small Regional Extent) is frequently used in the land-change science. The model has an integrated system-oriented approach that considers the land use system as a whole and thus includes a high degree of dynamics. The CLUE-s model is able to represent the temporal dynamics, competition between land use classes, and path-dependencies with possible irreversible consequences. The model is based on the system theory and allows the integrated analysis of land use change in the statistical relationship to socio-economical and biophysical driving forces. The CLUE-S model has been developed at the University of Wageningen, NL, under the direction of Peter Verburg. It is based on the former CLUE modelling approach, which has been developed for the national to continental scale (Veldkamp & Fresco 1996). The functionality of CLUE-s is explained in detail at Verburg et al. (2002). Previous fields of application concentrated i.a. on the future of agricultural areas in Europe (Verburg et al. 2006, 2008; Groot et al. 2009) and land use change in tropical regions, where the focus especially lay on the expansion of farmland account of tropical forests and savannahs. Examples for the adoption of the CLUE-s model for this kind of research can be found for South-East Asia (Verburg & Veldkamp 2004), the sub-Saharan Africa (Judex 2007), and South America (Wassenaar et al. 2007). Batisani and Yarnal (2009a, 2009b) concentrated on modelling urbanisation in North America with the CLUE-s model. Recently, the CLUE-s model is increasingly used in combination with other models. Hence, the simulated land use data are used e.g. as input for hydrological models (Dams, Woldeamlak & Batelaan 2008; Lin et al. 2008).

Figure 4: Concept of the Clue-S model. Modified and extended according to Verburg et al. (2002)

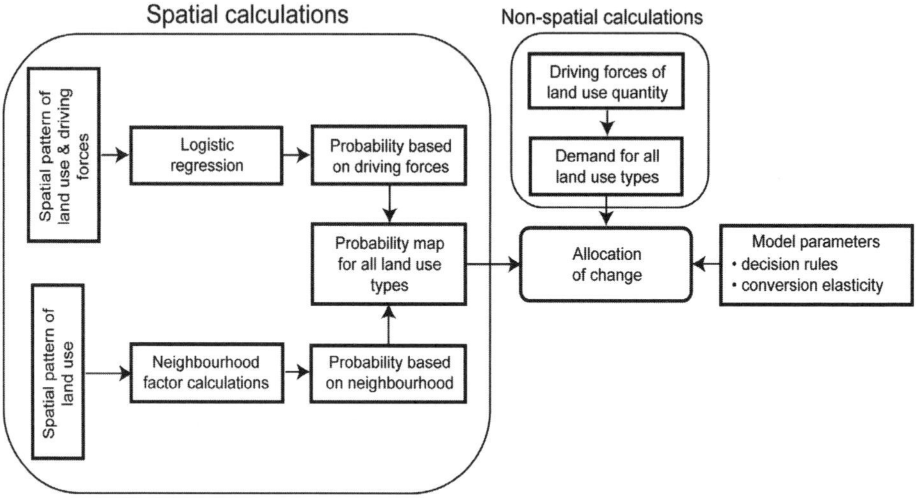

The core of the CLUE-s model consists of the allocation of land use changes, which depends heavily on scenario definitions (Figure 4). The definition of these scenarios and, therefore, the definition of the demand of land use is not part of the model and has to be added externally. The allocation of land use is carried out in the model by an iterative algorithm based on probability maps of the single land uses/land use changes. These probability maps are generated through a logistic regression analysis.

Furthermore, additional information is incorporated in the model, like conversion restrictions, neighbourhood weights, or the elasticity of the single land use classes. The elasticity is related to the reversibility of land use changes. That indicates high values for land use types with high capital investment, which will not easily be converted and low values for land use types that can easily be converted. The total probability Pt is calculated from the sum of the probabilities based on the driving forces $PA_{i,k}$ at each pixel i for every land use k, the neighbourhood probabilities $PN_{i,k}$, the conversion elasticity Ek, potential "location specific preference additions" (e.g. regional political promotion of certain processes) L_k, an iteration variable I_k and a very small random value r:

$$PT_{i,k} = PA_{i,k} + PN_{i,k} + E_k + L_k + I_k + r \quad (1)$$

In every step of the iteration, that land use class is assigned to the raster cell, which shows the highest total probability for that class (except exclusions based on the conversion matrix or protected areas). The amount of the allocated land use area per class is subsequently compared with the current demand of land use for that time step and I_k is increased or decreased in every iteration, until $PT_{i,k}$ fits to the demand within a range of tolerance.

A similar system-oriented model is, for example, the Land-Transformation Model, which is based on Neural Networks (Pijanowski et al. 2002), or GEOMOD, which considers neighbourhood relationships, spatial stratification and probabilities of change based on biophysical attributes. Though, this model on the one hand can only predict the location of change and not the magnitude of change and on the other hand can only model the change of a single land use category (Pontius et al. 2001).

Urban growth model based on Cellular Automata (SLEUTH)

Like the GEOMOD model discussed in the last section, most of the urban growth models are only able to model the change of one land use category, which is urban, and that mostly just in one direction, which is growth. One example is the SLEUTH model. SLEUTH is a CA-based urban growth and land use change model. The name stands for "**S**lope, **L**and use, **E**xclusion, **U**rban, **T**ransport, **H**illshade" and, thereby, comprises the major input parameters of the model. The model is explained in detail by Clarke et al. 1997. SLEUTH has been applied in a wide range of studies worldwide, with a focus on the USA (Yang & Lo 2003; Jantz et al. 2003; Xian & Crane 2005; Dietzel et al. 2005), due to the fact that the model has been optimised for urban growth processes in the US. However, there are a number of examples from Europe (Silva & Clarke 2005), South America or Southeast Asia (Lebel et al. 2007). The model consists of a core component, which simulates urban growth (the Urban Growth Model – UGM) and a Deltatron model, which considers the land use dynamic for other classes. Though, the former, which can be used separately, is used more frequently.

The SLEUTH model needs the topographical information slope and hillshade, although the latter is only needed for visualisation purposes. Consistent land use maps for two points in time have to be provided only for the land use Deltatron component, the UGM works without them. The exclusion layer determines, which areas in the study area cannot be changed (e.g. water bodies or protected areas) or, if not excluded, are by a certain degree resistant against urbanization. So the exclusion layer can influence the growth rate significantly. The urban areas are the most important drivers for the model. At last the model needs a transportation layer representing the road network in the study area.

Calibration is the most important step in every modelling application and requires the main focus of attention. In the calibration phase of SLEUTH a brute-force method is used in order to determine five calibration parameters. That means every possible combination of parameters is being tested until the best combination is found. In order to do this consistently and reproducible, every combination is tested with a specified amount of Monte-Carlo iterations. The five parameters control the transition rules that are implemented in the model, which in turn allow four different kinds of urban growth: spontaneous growth, diffusive (new spreading centres), organic (infill and edge growth) and road influenced growth. The parameters are:

Diffusion: Determines the dispersiveness of the outward distribution. It controls the number of pixels that are selected randomly for possible urbanisation.

Breed: The probability that a newly generated settlement starts its own growth.

Spread: Controls how much existing settlements radiate.

Slope resistance: Influences the likelihood of growth on steep slopes.

Road gravity: Influences the creation of new centres along roads.

In the study presented here we used a modified version of the UGM component of the SLEUTH model. It differs significantly in two points from the original model, which affects the model behaviour during the calibration period. Firstly, in the original model a set of Pearson's r^2 statistics is used to determine the growth coefficients, like the total amount of urban pixels, edges, clusters etc. Dietzel & Clarke (2007) describe how these statistics can be utilised to achieve optimal calibration results. We have substituted this procedure by a simple measure of the goodness of fit, the multiple resolution validation (MRV), which is described in the next chapter. Secondly, which is closely related to the first point, the original calibration procedure uses the least squares method to fit the simulated urban growth using the above mentioned statistics. This requires, beside an initial data set, at least three control data sets, which means the states of urban areas after the initial date. Four subsequent data sets and ideally a fifth one in order to validate the model are rarely available. Therefore, we use the MRV to compare the simulated built-up areas with only one control map during the process of model calibration.

How does the CA actually run in detail? Principally a certain number of cells in a continuous grid are selected by random and their properties and the properties of their neighbourhood are examined. The number and location of the randomly selected cells is controlled by the growth parameters. Depending on the type of growth different properties of the selected cells are looked at. For the diffusive growth (new spreading centres) e.g. this would be the existence of non-built-up cells in the direct proximity of a selected built-up cell and the slope of these cells. Depending on the specified parameters and transition rules the CA computes, if a cell is available for change or not. The CA knows only two states: 1 = urban/built-up and 0 = non-urban/non-built-up.

The modelling framework (XULU)

Models of land use change are highly complex applications that have been developed specially to meet the needs of modellers. Most of these models have been developed at universities and are a result of long-time research. Also from the point of software development the models match the tradition of the research groups (Table 1).

Upon these facts Schmitz et al. (2007) developed XULU (eXtendable Unified Land Use Modelling Platform), a modelling framework that carries out the task of all functionalities concerning input, output, editing and visualisation. Due to its modular architecture a specific model is implemented separately from the program itself (Figure 5). The core program provides data management (the so called data pool), I/O routines for data import and export, data structure, memory management and data visualisation. XULU is implemented in Java 1.6 and offers a model independent graphical user interface. The advantage of this software architecture is that in principle every model (not only land use models) can be implemented in XULU after translating the model algorithm into Java. Thus, the model developer does not need to worry about data and memory management any more. Moreover, for XULU it does not matter, if data is input or output, because all data is treated as a model resource. Consequently, the output of one model can serve as input for a second model both running parallel in

Table 1: Properties of selected land use change models

Model	Clue-S	SLEUTH	GEOMOD	Land-Transformation Model
Source Code	C++	C++	Part of the Idrisi Software	C, Avenue
OS	Windows	Unix	Windows	Windows
Data format	ASCII ArcGrid	GIF	RST (Idrisi Raster Format)	ASCII ArcGrid
Visualisation	No	Yes	Yes (in Idrisi)	No
Naming conventions	e.g. Cov1_0.0, Sc1gr0.fil	<location>.urban.<year>.gif	-	-
GUI	Yes	No	Yes (Idrisi)	No

Figure 5: Separation of the XULU modelling framework from a certain model implementation. Modified according to Judex (2007).

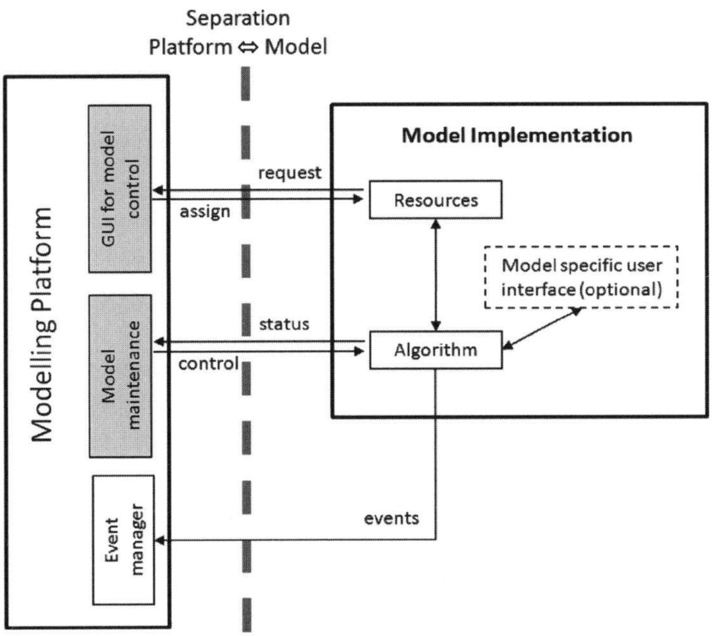

XULU. So the detour via GIS, where data of one format has to be transferred into another format, can be omitted. Using the common framework, model calibration and validation needs to be implemented only once for all models, which facilitates inter-model comparison.

Calibration and validation strategy

There is no consistent agreement in land use modelling about model comparison or even on how different model runs should be evaluated. Amongst others, this is due to diverging objectives that model users have. While one group wants to make predictions as accurate as possible, the other group emphasises the ability of a model to support the general knowledge of processes and mechanisms of land use change (Pontius Jr & Malanson 2005). In the so far most extensive comparison of land use models (Pontius Jr et al. 2008), seven laboratories with 13 applications, nine different models and 12 study areas participated. As one statistical measure of validity, the method of multiple resolution validation (MRV) was used, which has been implemented as a separate module in XULU. Thanks to this comprehensive work, a model comparison that uses a consistent method for model validation is available for the first time. Because of that, this method was chosen to evaluate the model runs presented in this study. The MRV takes the quantity as well as the location of change into account. This method is described in detail for the comparison of categorical land use data by Costanza (1989) and Pontius et al. (2004). Usually maps are compared pixel-wise, e.g. with the Kappa coefficient. This causes a problem for land use modelling, because models rarely hit the exact position of real change, but very well account for representing patterns of change. At a pixel-by-pixel analysis every pixel is counted as error, where the model map does not exactly fit the reference map, no matter if the pixel is located in the neighbourhood. With the MRV four neighbouring pixels are averaged stepwise. So with every step the length of the pixel side increases by the factor of two. The amount of correct pixels increases in every step, until the last step, where the whole study area is inside of one big pixel and both location agreement and location disagreement approach 0.

Because a study area is usually not perfectly square, every cell has to be weighted according to its fraction in the larger context. The weight *Wn* results from the number of smaller cells that make up a larger cell. Equation 1 shows the agreement *Fg* between the maps *R* (reference) and *S* (simulation) at resolution *g*, where every cell *n* has the weigh *Wn*. *Rnj* (*Snj*) is the fraction of class *j* in grid cell *n* of the reference map *R* (or simulation map *S*) and *Ng* is the number of grid cells in the given map resolution. With this equation the overall agreement for every resolution can be calculated (Pontius Jr. 2002):

$$F_g = \frac{\sum_{n=1}^{Ng} \left[Wn \sum_{j=1}^{J} MIN(Rnj, Snj) \right]}{\sum_{n=1}^{Ng} Wn} \qquad (2)$$

Furthermore, it is helpful to look at a measure of agreement that combines all resolutions. Therefore, we use the weighted average of each resolution, so that higher resolutions gain a higher weight and coarse resolutions a lower one, but without ignoring them. Here we use the following equation taken from Costanza (1989):

$$F_t = \frac{\sum_{g=1}^{n} F_g e^{k(g-1)}}{\sum_{g=1}^{n} e^{k(g-1)}} \qquad (3)$$

whereby F_t is the weighted average over all resolutions, F_g the measure of agreement at resolution g, and k a constant. Herewith exponentially less weight is given to the agreement of coarse resolutions. The constant k gives subjectively more weight to the higher or coarser resolution. If $k = 0$ all resolutions have the same weight, if $k = 1$ only the first resolution levels are considered. To assess land use patterns Costanza (1989) proposes a value of $k = 0.1$, which we retained.

In order to evaluate model results with the method described above, three data sets are necessary: (1) a reference map of time 1, (2) a reference map of time 2 and (3) a simulation map of time 2. The reference map of time 1 is the initial point for modeling. At the same time it serves as a Null-model, which is the assumption that no change has taken place. Therefore, the reference maps of t1 and t2 are compared. To evaluate the model result, the simulation map of t2 is compared with the reference map of t2. Figure 6 shows a part of the study area and the observed change between t1 and t2 (A) as well as the simulated change between t1 and t2 (B).

In Figure 7 one can check visually which errors have occurred during the simulation. For Figure 7 (B) all three input data sets have been used. With the MRV the fraction of correct pixels is computed for every resolution with these three maps. The coarser the resolution of the land use change model and the Null-model the higher the agreement, because the error that results from the inaccurate location of the simulated change is being minimised.

Figure 8 shows the first six steps of the series of coarsening resolutions at the example of the Cologne/Bonn region with the result of the simulation with the CLUE-s model of built-up areas for 2001 based on 1984.

In most cases, land use models are less accurate at a high resolution than the corresponding Null-model (Pontius Jr. et al. 2008). Usually, at a certain resolution the accuracy of the land use model exceeds the accuracy of the

Figure 6: The Cologne/Bonn region in the southern part of NRW: (A) observed change of built-up areas between 1984 and 2001, (B) simulated change of built-up areas between 1984 and 2001 with CLUE-s

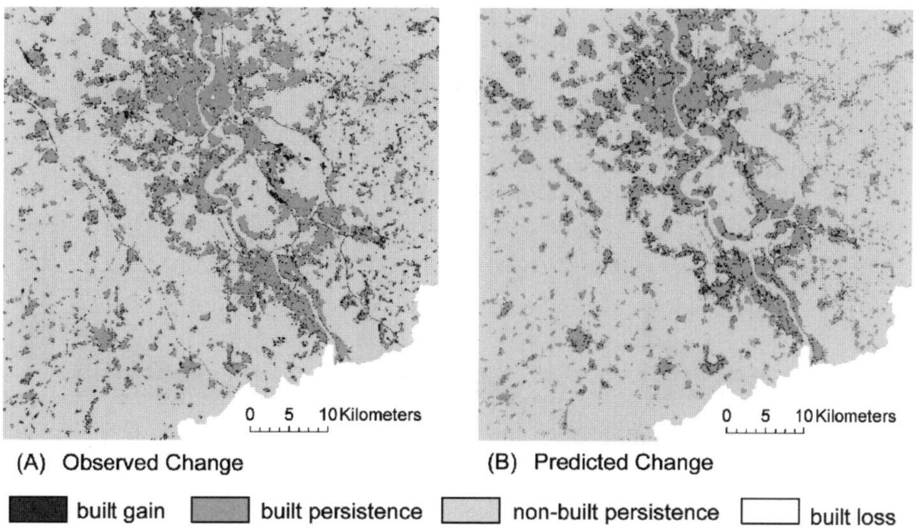

Figure 7: Cologne/Bonn region in the south of NRW: (A) Error in the simulation map 2001, (B) overlay of reference map 1984, reference map 2001 and simulation map 2001 (CLUE-s)

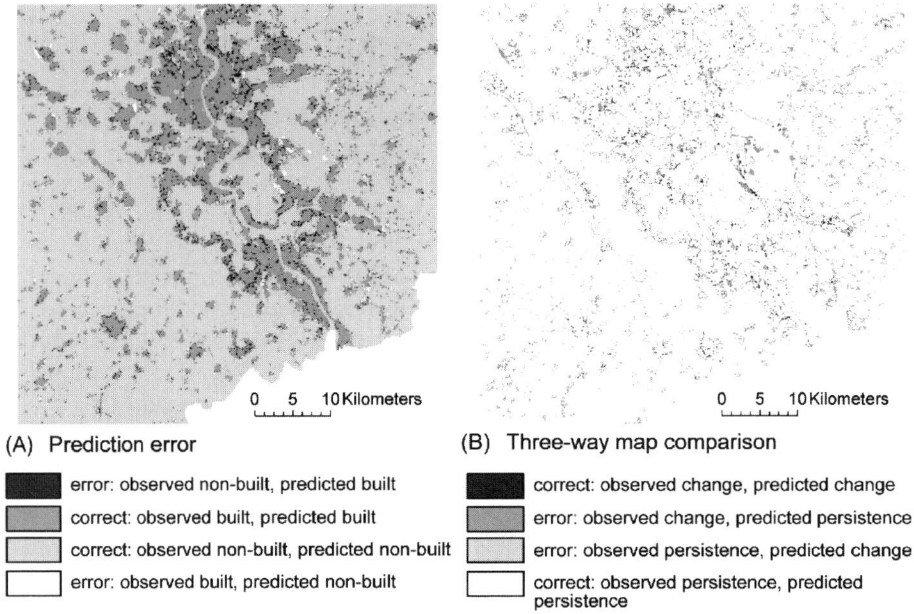

Figure 8: Agreement between the reference map 2001 and the simulation of 2001 (CLUE-s) for the Cologne/Bonn region at different resolutions. The numbers are multiples of the pixel size.

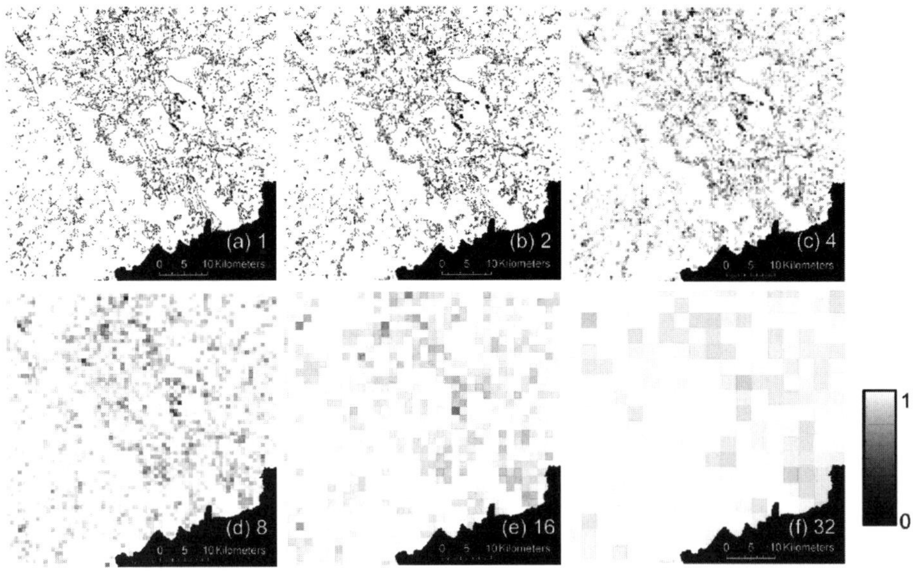

Null-model, which is called the Null-resolution (Pontius Jr. et al 2004). A lower Null-resolution indicates that location errors occur at small distances. Figure 9 gives an example of the accuracy development for the simulation of the built-up class with the CLUE-s model in NRW at different resolutions in comparison to the Null-model.

Figure 9: Fraction correct at different resolutions for the simulation of 2001 with the CLUE-s model in relation to the Null-model

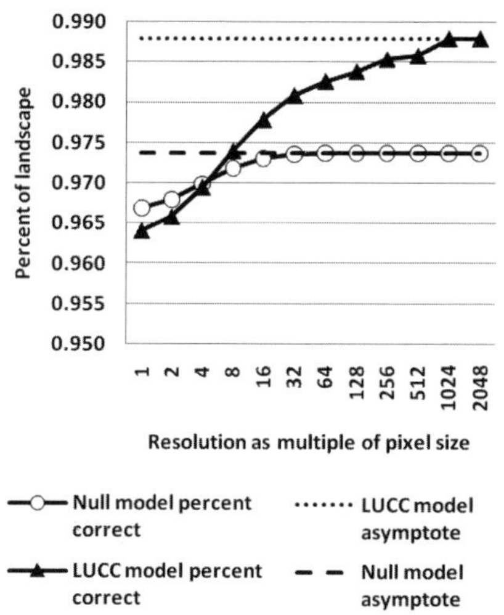

The MRV method has been used to achieve the optimal parameterisation for the models during the calibration phase as well as for the validation of the model results. In many papers there is no clear distinction between calibration and validation, which is criticised by Pontius et al. (2004). Oftentimes, there is no clear explanation if there is a model fitting or a real prediction and sometimes it is just the lack of available data to perform a validation. Especially because land use models are used to predict future changes of land use, they are understandably difficult to validate.

In our work we made the effort to distinct calibration and validation procedures. Both models have been calibrated based on the year 1984 to model the year 2001. In case of the CLUE-S model also information of 2001 was incorporated. Firstly, the observed land use demand for 2001 was specified and for the logistic regression we did not use the class "built-up" as the dependent variable, but the positive change of built-up areas between 1984 and 2001. The prediction has been calculated for the years 2001 to 2025. A validation of the 2025 results was, of course, not possible, so we used the land use classification of 2005 for validation. Due to the fact that land use normally does not change linearly, it is to a certain extent problematic that the time frame for validation is much shorter than the one for the calibration. Additionally, the observed change in that short time period has been quite low and so most of the land use is persistent. For a better comparison of calibration and validation results we used the MRV for both modelling steps as a measure of accuracy.

Structural comparison of the models

The CLUE-s model and the urban growth model based on SLEUTH, which have both been implemented in the XULU modelling framework show different structural features, of which some are outlined in Table 2.

The major differences are that CLUE-s is able to model different concurrent land use classes, while the Urban Growth Model knows only one direction for one land use class. CLUE-s has the more substantial theoretical and methodical background, but does not distinguish between land use classes in the manner of modelling. The overall spatial probability distribution, as calculated with Equation 1, is always critical for the spatial allocation of a land use class. In the case of modelling built-up land with CLUE-s, areas with a high probability are usually close to existing agglomerations and are hence preferred

Table 2: Selected characteristics of the Clue-S model and the CA Urban Growth Model (SLEUTH)

Characteristic	Clue-S	CA Urban Growth Model (SLEUTH)
1) Number of Land use categories	2 or more	2 (urban / non-urban)
2) Transition	Growth and shrinkage	Growth
3) Information about amount of change	Land use demand has to be calculated externally (e.g. with an economic model)	No information; is adjusted during the calibration process based on a reference map
4) Land use maps needed	At least one and second for calibration	At least one and a second for calibration (orig. SLEUTH needs 5 maps of urban extent for calibration)
5) Probability maps	Are created during model run with a logistic regression	None

for the allocation of more built-up land by the model. But in reality settlements can grow in urban and in rural areas as well with the approximately same rate. That is due to the fact that different driving forces are responsbile for the same land cover change (growth of built-up areas). However, the SLEUTH model and the modified Urban Growth Model used in this study comprise certain theoretical assumptions about the way settlements grow (edge growth, new spreading centres and road influenced growth).

Results

Calibration procedure

Figure 6 already showed the calibration results of the CLUE-s model for a subset of the study area. The result shows that Clue-S modelled at a spatial resolution of 100 m 96.4 % of the built-up areas in 2001 correctly based on the initial map of 1984 with the parameters used. At the same spatial resolution the Null-model shows an agreement of 96.7 %, which means that the prediction of no change would be slightly more accurate than the application of CLUE-s, because of the high persistence of built-up areas. However, Figure 9 shows that from the Null-resolution of 400 m CLUE-s performs better than the Null-model. The total weighted measure of agreement *Ft* results for CLUE-s in an agreement of 96.8 % and for the Null-model in an agreement of 96.9 %. This value can be taken as the goodness of fit for the model fitting. Thus, the CLUE-s model is not able to depict the change in built-up land better than the Null-model. Due to the fact that CLUE-s does not model change in built-up ar as alone, but the land use system as a whole by the competition of 6 land use categories, it is worth a look at the accuracy of the complete model run. There, the Null-model achieves a value of Ft = 0.887. So 88.7 % of the land use configuration between 1984 and 2001 stayed unchanged. With the CLUE-s model we achieved an agreement of 89.1 %. The model calculates the change of the whole land use configuration better than the Null-model.

The calibration of the UGM results in an agreement of 96.5 % for the built-up/non-built-up categories between the reference map of 2001 and the map of 2001 fitted with the model at 100 m resolution (Null-model: 96.7 %). The weighted average of all resolutions *Ft* shows that 97.0 % (Null-model 96.9 %) have been modelled correctly. Figure 10 gives a look at the goodness of fit for the calibrated models. While the UGM achieves at 2x the length of the pixel side the Null-resolution, it does not improve after a coarser resolution than 64-times the length of the pixel side. Clue-S represents the change in built-up land worse at higher resolutions than the Null-model, but increases constantly at coarser resolutions.

Figure 10: Goodness of fit for the calibration of two models for 2 land use categories (built-up / non-built-up) in relation to the Null-model with the MRV

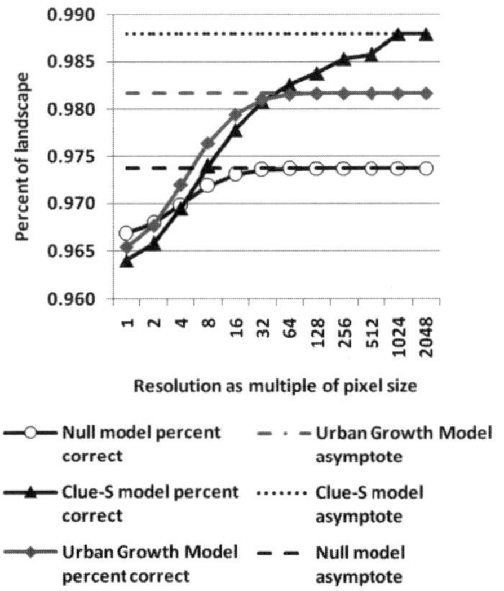

Both models have been calibrated in order to represent land use change in NRW better than the Null-model. For CLUE-s this does not apply to the binary classification built-up/non-built-up, but for the whole land use system. The UGM is able to predict urban growth better than the Null-model.

Validation procedure

If one wants to persuade land managers of a model, it is not sufficient to deliver a well calibrated model. In the presented case, the measure of the goodness of fit with the MRV for the data set of 2001 is not enough, because information of this data set has been incorporated in the model fitting, and thus, this data set is not completely independent. Thus, it is essential to validate the model with an independent data set. Therefore, we carried out the prediction of land use in 2025, starting with the year 2001, leaving all parameters identical to the calibration phase. The expansion of built-up land should be modelled with the UGM with the same rate as between 1984 and 2001. For CLUE-s the land use demand has been linearly extrapolated. For validation we used a land use classification of 2005. That means that an intermediate result of the model run is validated.

The weighted average of the MRV for all resolutions shows for the Null-model and the built-up/non-built-up binary map an agreement of 99.6 %, for the ClUE-s model 99.5 % and for the UGM 99.5 % as well. Those values are extraordinary high and very close to each other, which is primarily the result of the fact that the observed change in built-up land for the time span of only 4 years is quite small, namely 5264 pixels between 2001 and 2005. Thus, the validation result is dominated by the persistence of land use. In order to achieve an appropriate validation result, a similar time for the simulation would be needed like in the calibration phase of 1984-2001. At a small growth of 0.15 % of the total land area a model is hardly able to predict the location of change at a desired accuracy. But it is worth a look at the MRV results, as depicted in Figure 11.

Figure 11: Validation of model results (CLUE-s and UGM) for the binary built-up/non-built up classification for 2005 at different spatial resolutions

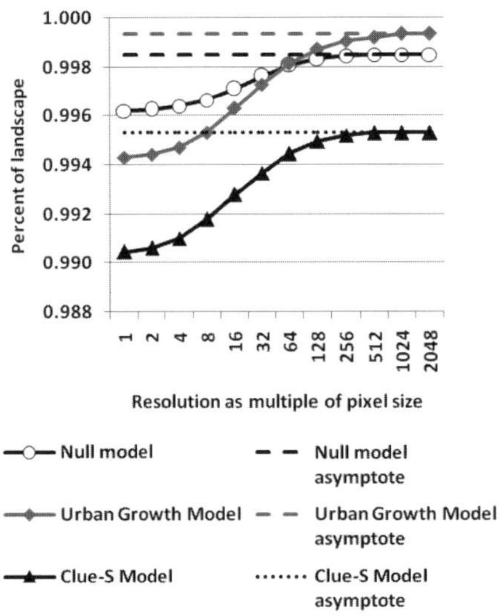

With a coarsening resolution the Null-model gains weakly in accuracy. The allocation of new built-up pixels with the CLUE-s model does not seem to contribute properly, because no Null-resolution is being achieved. Here, the assumption of no change would be more accurate. Considering the whole land use configuration, the value of Ft = 0.975 is quite good compared to the Null-model (Ft = 0.976), but here the curve also runs parallel to the one of the Null-model, and finally achieves the Null-resolution between 512 and 1024-times the length of the pixel side. The UGM at least achieves the Null-resolution at 64-times the original resolution. Altogether, it has to be noted that the behaviour of the UGM can reasonably confirm the calibration in a validation after 4 years of model run, while CLUE-s did not simulate the change in built-up land sufficiently correct.

Discussion

A comparison of two land use change models was carried out to test the performance and capability of two models with different theoretical and methodological approaches. The CLUE-s and the SLEUTH model approaches are calibrated to predict the urban land use changes and in the case of the CLUE-s approach other land use classes as well.

The comparison of land use change models cannot lead to the conclusion that one model is better than the other. It is always important to keep track of the aim of modelling. Depending on the model comparison we showed here, the conclusion can be drawn that for the simulation of urban growth in Central Europe an urban growth model, like the SLEUTH model, provides better results than a statistical system-oriented model like CLUE-s. In order to simulate changes in the whole land use system, an urban growth model, of course, is not sufficient and a more complex model, like CLUE-s, should be considered. If a model is to be handed over to land managers, it is not only necessary to proof comprehensible that a model is well calibrated, which is the case for the two presented models. Also a measure for the goodness of fit should be delivered that goes beyond calibration. So a model has to be validated. At this point, often difficulties arise, as it could also be shown in this study. A scientist always seeks to be successful, what often leads to the neglect of model validation, because this might unmask years of work as insufficient. Pontius et al. (2004) discuss this point in detail and give promising suggestions. Above all, it is important to explain the model user what potential suboptimal validation results might be concerned with. In our case, this in the first instance is the small observed land use change in the short validation period. This problem has also been noticed in the model comparison by Pontius et al. (Pontius Jr. et al. 2008), in which those models showed the lowest predictability that had the smallest net change. Another point which is discussed in this paper and that became obvious by applying the two presented models is that the modeled quantity of change is often smaller than the observed change. This is related to the fact that a model takes the risk of predicting an error at every simulated change. Thus, the model calibration often results in predicting less change, whereby potential errors are minimized. When a stronger change has taken place according to the reference map, a larger statistical signal is offered to the model during calibration.

Another point that potentially leads to a weakness in model validation is the non-linearity of land use systems. In the present example a linear development has been assumed during calibration. But the development of land use demand is possibly based upon an exponential or sigmoidal trend, which is not apparent when looking only at two steps in time. Furthermore, a change in development based on external factors can slow down or speed up land use change, which has not been apparent in the calibration phase.

At last, the accuracy of the reference map can lead to weaknesses in land use models. In the present example the accuracy of the land use classifications has been about 86-90 %, which is quite good for remote sensing products. But when a potential error of about 10 % faces a change rate of the same amount, the conclusion can be drawn that a part of the measured change is not real but erroneous and that a part of the measured persistence is change in real.

Land use change is a dynamic process. Feedbacks and path-dependencies are typical for every land use system. But if those are not known or not completely understood, we first assume a linear development of the model. The application of a model should always be associated with expert knowledge which can absorb potential weaknesses. So, after calibration and validation the model user should define scenarios of land use change, that incorporate explicit expert knowledge. In those scenarios it can be defined, if changes occur faster or slower than assumed for the calibration period. Furthermore, by assuming a statistically assured linear development as a basis, possible scenarios can be compared to that.

In this study the XULU modeling framework has been proven to be a good option in order to compare models of land use change. Both models had access to the same data pool and could be calibrated and validated with the same tool, the MRV, which has been implemented in XULU explicitly for this purpose.

Annotations

The land use classifications of the reference maps have been created within the project "Visualisierung von

Landnutzung und Flächenverbrauch in Nordrhein-Westfalen mittels Satelliten- und Luftbildern", funded by the Ministry of Environment, Nature Conservation, Agriculture and Consumer Protection, NRW (grant no. 01 LW 06001B; http://www.flaechennutzung.nrw.de).

The modelling platform XULU has been developed within the Impetus project, funded by the German Federal Ministry of Education and Research and the Ministry of Innovation, Science, Research and Technology, NRW (grant no. 313-21200200) and is continued as an Open Source software under the GNU-LGPL license. We express our gratitude to the initiators of XULU, Hans-Peter Thamm and Thomas Bode, as well as to the developers Martin Schmitz, Dominik Appl and Stefan A. Krüger.

References

Antrop, M. (2004): Landscape change and the urbanization process in Europe. In: Landscape and Urban Planning, vol. 67, pp. 9-26.

Batisani, N. & B. Yarnal (2009a): Uncertainty awareness in urban sprawl simulations: Lessons from a small US metropolitan region. In: Land Use Policy, vol. 26, no. 2, pp. 178-185.

Batisani, N. & B. Yarnal (2009b): Urban expansion in Centre County, Pennsylvania: Spatial dynamics and landscape transformations. In: Applied Geography, vol. 29, no. 2, pp. 235-249.

Briassoulis, H. (2000): Analysis of Land Use Change: Theoretical and Modeling Approaches. Regional Research Institute, University of West Virginia. Retrieved from http://www.rri.wvu.edu/WebBook/Briassoulis/contents.htm

Clarke, K., Hoppen, S. & L. Gaydos (1997: A self-modifying cellular automaton model of historical urbanization in the San Francisco Bay area. In: Environment and Planning B: Planning and Design, vol. 24, no. 2, pp. 247-261.

Costanza, R. (1989): Model goodness of fit: A multiple resolution procedure. In: Ecological Modelling, vol. 47, no. 3-4, pp. 199-215.

Dams, J., Woldeamlak, S. & O. Batelaan (2008): Predicting land use change and its impact on the groundwater system of the Kleine Nete catchment, Belgium. In: Hydrology and Earth System Sciences, vol. 12, pp. 1369-1385.

Dietzel, C. & K. Clarke (2007): Toward Optimal Calibration of the SLEUTH Land Use Change Model. In: Transactions in GIS, vol. 11, no. 1, pp. 29-45.

Dietzel, C., Herold, M., Hemphill, J. & K. Clarke (2005): Spatio-temporal dynamics in California's Central Valley: Empirical links to urban theory. International Journal of Geographical Information Science, vol. 19, no. 2, pp. 175-195.

European Environment Agency (2007): Land use scenarios for Europe: qualitative and quantitative analysis on a European scale, European Environment Agency, Copenhagen.

Goetzke, R., Over, M. & M. Braun (2006): A Method to map Land use Change and Urban Growth in North Rhine-Westphalia (Germany). In: Proc. 2nd Workshop of the EARSeL SIG on Land Use and Land Cover, Bonn, Germany, pp. 102-111.

Groot, J., Rossing, W., Tichit, M. & N. Turpin (2009): On the contribution of modelling to multifunctional agriculture: Learning from comparisons. In: Journal of Environmental Management, vol. 90, no. 2, pp. 147-160.

Jantz, C., Goetz, S. & M. Shelley (2003): Using the SLEUTH urban growth model to simulate the impacts of future policy scenarios on urban land use in the Baltimore / Washington metropolitan area.

In: Environment and Planning B: Planning and Design, vol. 31, pp. 251-271.

Judex, M. (2008): Modellierung der Landnutzung in Zentralbenin mit dem XULU-Framework. Diss., Bonn, Germany.

Lambin, E. (2004): Modelling land use change. In: Wainwright, J. & M. Mulligan (eds.): Environmental Modelling. Finding Simplicity in Complexity, John Wiley & Sons, Chichester , pp. 245-254.

Lebel, L., Thaitakoo, D., Sangawongse, S. & D. Huaisai (2007): Views of Chiang Mai: The Contribution of Remote-Sensing to Urban Governance and Sustainability. In: Netzband, M., Stefanov, W. & C. Redman (eds): Applied Remote Sensing for Urban Planning, Governance and Sustainability, Springer Verlag, Berlin, pp. 221-247.

Lin, Y., Lin, Y., Wang, Y. & N. Hong (2008): Monitoring and Predicting Land use Changes and the Hydrology of the Urbanized Paochiao Watershed in Taiwan Using Remote Sensing Data, Urban Growth Models and a Hydrological Model. In: Sensors, vol. 8, pp. 658-680.

Mulligan, M. & J. Wainwright (2004): Modelling and model building. In: Wainwright, J. & M. Mulligan (eds): Environmental Modelling. Finding Simplicity in Complexity, John Wiley & Sons, Chichester, pp. 5-74.

Pijanowski, B., Brown, D., Shellito, B. & G. Manik (2002): Using neural networks and GIS to forecast land use changes: a Land Transformation Model. In: Computers, Environment and Urban Systems, vol. 26, no. 6, pp. 553-575.

Pontius Jr, R. (2002): Statistical Methods to Partition Effects of Quantity and Location During Comparison of Categorical Maps at Multiple Resolutions. In: Photogrammetric Engineering & Remote Sensing, vol.68, no. 10, pp. 1041-1049.

Pontius Jr, R., Boersma, W., Castella, J., K. Clarke (2008): Comparing the input, output, and validation maps for several models of land change. In: Annals of Regional Science, vol. 42, pp. 11-37.

Pontius Jr, R., Cornell, J. & C. Hall (2001): Modeling the spatial pattern of land use change with GEOMOD2: application and validation for Costa Rica. In: Agriculture, Ecosystems and Environment, vol. 85, pp. 191-203.

Pontius Jr, R., Huffaker, D. & K. Denman (2004): Useful techniques of validation for spatially explicit land-change models. In: Ecological Modelling, vol. 179, no. 4, pp. 445-461.

Pontius Jr, R. & J. Malanson (2005): Comparison of the structure and accuracy of two land change models. In: International Journal of Geographical Information Science, vol. 19, no. 2, pp. 243-265.

Pontius Jr, R., Shusas, E. & M. McEachern (2004): Detecting important categorical land changes while accounting for persistence. In: Agriculture, Ecosystems and Environment, vol. 101, pp. 251-268.

Schmitz, M., Bode, T., Thamm, H. & A. Cremers (2007): XULU - A generic JAVA-based platform to simulate land use and land cover change (LUCC). In: Oxley, L. & D. Kulasiri (eds): Proceedings MODSIM 2007 International Congress on Modelling and Simulation, Modelling and Simulation Society of Australia and New Zealand, pp. 2645-2649.

Schöttker, B., Over, M., Braun, M., Menz, G. & A. Siegmund (2004): Monitoring state-wide urban Development using multitemporal, multisensoral Satellite Date covering a 40 Year-Time Span in North Rhine-Westphalia (Germany). In: Proceedings SPIE 2003, Barcelona, Spain, pp. 252-261.

Silva, E. & K. Clarke (2005): Complexity, Emergence and Cellular Urban Models: Lessons Learned from Applying Sleuth to Two Portuguese Metropolitan Areas. In: European Planning Studies, vol. 13, no. 1, pp. 93-115.

Veldkamp, A. & L. Fresco (1996): CLUE: a conceptual model to study the Conversion of Land Use and its Effects. In: Ecological Modelling, vol. 85, no. 2-3, pp. 253-270.

Verburg, P., Eickhout, B. & H. van Meijl (2008): A multi-scale, multi-model approach for analyzing the future dynamics of European land use. In: Annals of Regional Science, vol. 42, pp. 57-77.

Verburg, P., de Groot, W. & A. Veldkamp (2003): Methodology for multi-scale land use change modelling: concepts and challenges. In: Dolman, A., Verhagen, A. & C. Rovers (eds): Global Environmental Change and Land Use, Kluwer Academic Publishers, Dordrecht, NL, pp. 17-52.

Verburg, P., Kok, K., Pontius Jr, R. & A. Veldkamp (2006): Modelling Land use and Land-Cover Change. In: Lambin, E. & H. Geist (eds): Land use and Land-Cover Change. Local Processes and Global Impacts, Springer, Berlin, pp. 117-136.

Verburg, P., Schot, P., Dijst, M. & A. Veldkamp (2004): Land use change modelling: current practice and research priorities. In: GeoJournal, vol. 61, pp. 309-324.

Verburg, P., Soepboer, W., Veldkamp, A. & R. Limpiada (2002): Modelling the Spatial Dynamics of Regional Land Use: The CLUE-S Model. In: Environmental Management, vol. 30, no. 3, pp. 391-405.

Verburg, P. & A. Veldkamp (2004): Projecting land use transitions at forest fringes in the Philippines at two spatial scales. In: Landscape Ecology, vol. 19, pp. 77-98.

Verburg, P., Veldkamp, A. & M. Rounsevell (2006): Scenario-based studies of future land use in Europe. In: Agriculture, Ecosystems and Environment, vol. 114, pp. 1-6.

Vitousek, P., Mooney, H., Lubchenco, J. & J. Melillo (1997): Human Domination of Earth's Ecosystems. In: Science, vol. 277, pp. 494-499.

Wassenaar, T., Gerber, P., Verburg, P. Rosales , M et al. (2007): Projecting land use changes in the Neotropics: The geography of pasture expansion into forest. In: Global Environmental Change, vol. 17, no. 1, pp. 86-104.

Xian, G. & M. Crane (2005): Assessment of urban growth in the Tampa Bay watershed using remote sensing data. In: Remote Sensing of Environment, vol. 97, pp. 203-215.

Yang, X. & C. Lo (2003): Modelling urban growth and landscape changes in the Atlanta metropolitan area. In: International Journal of Geographical Information Science, vol. 17, no. 5, pp. 463-488.

Value Orientation, Locational Choice and Residential Satisfaction – a Spatial Econometric Analysis

Christina West and Philipp Deschermeier

Introduction

Efficient urban development strongly requires information about the actual and future residential needs and requests. Without sufficient knowledge about the reasons why inhabitants are moving from one district to another and what determines their satisfaction level at their actual urban residence, the exodus of residents of towns like Mannheim (307,100 inhabitants at the end of year 2005) towards the suburban hinterland cannot be stopped. But also for foreign people, who are moving to the city from other towns or states, it would be a great help to know beforehand the districts or suburbs which suit them best, so that they do not have to move again after getting to know the different residential zones.

Residential satisfaction and persistency are mainly governed by the hierarchical realisation of needs in the present neighbourhood or adjacent districts. Depending on factors like age, sex, position in the live cycle, living arrangements, education, profession or ethnic background, different needs are more or less important. Thus, as a first hypothesis, the motivation for changing residential locations is primarily expected to be influenced by the basic conduct of life and the corresponding individual preferences (cp. West 2007, West et al. 2008, see also Meng et al. 2008 and Gans et al. 2009). These preferences are influenced by the individual fundamental value orientations of the migrants and intra-urban movers. Therefore, and this is the second hypothesis, the assessment of these fundamental value orientations is the key for studying and understanding the motivation for migrations between city and suburban surroundings as well as removals between different urban districts, and consequently for answering the question "Who is living where and why?".

In the present paper, this question is addressed to the residents of the 17 districts of Mannheim/Germany. As a first step, fundamental value orientations are constructed on the basis of data from a survey carried out for the City of Mannheim and the GBG – Mannheimer Wohnungsbaugesellschaft mbH (Municipal Housing Enterprise) in 2007 and 2008 using exploratory factorial analysis. A subsequent typification of the interrogated residents by cluster analysis leads to eight value orientation groups which can be assumed to have similar fundamental value orientations and, thus, comparable residential needs (ch. 2; 3).

In a second step, the residential satisfaction, which also has been assessed in the survey, is modelled for the different value orientation groups as a function of their position with respect to the fundamental value orientation factors. Two types of econometric models are used which include spill-over effects between neighbouring districts.

Migration Study in Mannheim – Data

The background to the present analysis is a migration study which was carried out on request of the City of Mannheim and the GBG – Mannheimer Wohnungsbaugesellschaft mbH (Municipal Housing Enterprise) in 2007 and 2008 (West et al. 2008; Gans et al. 2009). The City of Mannheim made a data set available comprising age, sex and nationality of all people who changed residence between 1st January 2006 and 30th June 2007, including removals inside Mannheim, out-migration from and in-migration to Mannheim. From this record, a data subset was extracted according to two criteria: the people surveyed were required to be at least 18 years old and to live or have lived in Mannheim or in the commuting hinterland of the city. In total, three double-stratified random samples for a written survey were drawn from people who had performed a change of residence with destination Mannheim. The response rate was about 16 per cent in each sample which lead to a total number of 1,188 households for analysis. The questionnaire addresses the household and individual characteristics of the 2,508 surveyed household members, such as sex, age, nationality, educational level and occupation, reasons for the move, housing needs and aspirations as well as value orientations. Additionally, a double-stratified random sample of people who did not move over the past ten years was drawn for a written survey with comparable questions. In this sample, the response rate was about 20 per cent providing data of a total of 315 households with 756 members. Statistical tests concerning the age distribution in the total population and in each sample show no significant deviations on the level $\alpha = 0.05$.

The representativeness of the data is restricted in two respects: a comparison with secondary statistics from the reporting authorities of Mannheim shows that, on the one hand, persons between 18 and below 25 years old are underrepresented, especially in the out-migration sample. As a consequence, persons who are between 30 and below 50 years old, i.e. the group of potential suburbanisers, are partly over-represented. On the other hand, persons with non-German nationality or with a migration background are underrepresented.

Moreover, two internet-based surveys were added to supplement the results of the written survey: the important objective was to interrogate white collar workers of enterprises located in Mannheim about their residential motivations. A total of 661 employees participated in the online survey. Half of them lived outside Mannheim. As prospective professionals and managers, students of universities and colleges in Mannheim, they form an important group in terms of assessing the city as a future residential and working location. A total of 383 students participated in the online survey.

The questionnaire was designed with the intention of constructing fundamental value orientations.

Fundamental value orientation and place of residence

The motivation for changing residence is influenced by the conduct of life and, herewith, by individual preferences. Therefore, a key question of this analysis was to find out how the fundamental intrinsic motivation is developed, which is the basis for the decisions and actions leading to the preferences and to the context for choosing a place of residence, i.e., how subjective experience leads to subjective expectations through subjective processing. These experiences and expectations determine the activity potentials which play a major role for the constitution of inter-individual and different conducts of life. Despite the equal underlying circumstances of existence or lifestyles (social layers), they lead to "subtle distinctions" (Bourdieu 1982).

In a first step, fundamental value dimensions were constructed. On this basis, all respondents from the postal and online surveys now living or having lived in Mannheim, aggregated into 17 districts, could be characterised. Comparable data from all respondent types suitable for reflecting fundamental value dimensions were selected. The variables of the four main subject areas, namely social life and gender role, potential of activity, general preferences with respect to the residential area and the socio-economic profile depending on cultural and economic resources, were analysed separately via an explorative factorial analysis (EFA) with the aim of identifying the dimensionality of the particular subject area. All in all, seven dimensions in four

Figure 1: Dimensions of fundamental value orientations (source: own analysis)

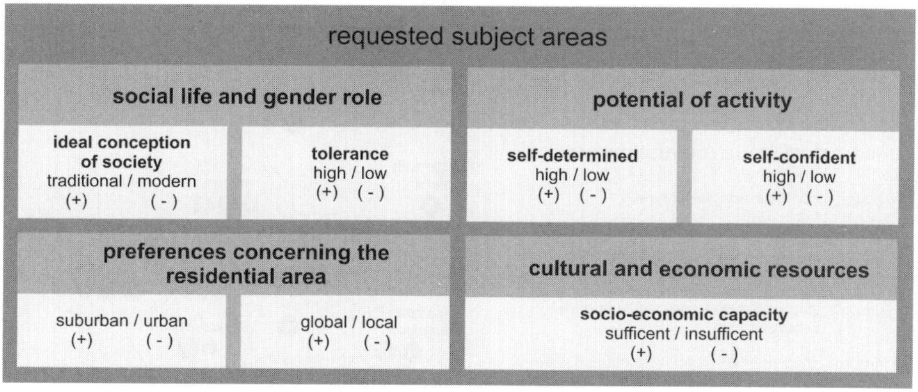

Table 1: Cluster centres of the value orientation groups (source: own analysis)

value orientation group	1	2	3	4	5	6	7	8
conception of society (+)/(-)	-0.29	-0.16	-0.20	-0.37	0.08	0.95	-1.03	1.20
tolerance (+)/(-)	0.32	0.37	0.58	0.23	-0.33	0.63	-1.24	-1.14
self-determined (+)/(-)	0.09	0.47	0.45	-0.10	-0.44	-1.33	0.58	0.00
self-confident (+)/(-)	0.18	0.23	0.16	0.12	-1.57	0.60	0.11	0.10
suburban/urban (+)/(-)	-0.05	0.06	0.62	-1.61	-0.08	0.62	0.32	0.04
global/local (+)/(-)	-1.39	0.21	0.39	0.55	-0.14	0.69	-0.50	0.17
socio-economic capacity (+)/(-)	0.22	1.42	-0.59	-0.47	-0.27	-0.47	0.42	-0.26

surveyed thematic areas were identified and quantified as indices by extraction of the corresponding factors (Figure 1). The designations of the dimensions and their dichotomous values corresponding to positive (+) or negative (-) factor values are given in figure 1.

In a second step, on the basis of the seven dimensions of the fundamental value orientations (figure 1), the respondents were subdivided using cluster analysis, resulting in eight value orientation groups with similar characteristics (table 1). The groups reflect different requirements concerning the residential area and the residence itself and are characterised by an archetype resident (e. g. "elderly passive type"). The distribution of these groups in the districts of Mannheim is shown in figure 2, where the groups with higher-than-average and strongly higher-than-average abundance are indicated by small and big circular symbols. It is remarkable that the "student type" is mainly located in the districts in or near the inner city, while the "domestic peripheral type" prefers the peripheral districts.

The main focus of this paper is placed on the observed average residential satisfaction for the 17 districts of Mannheim and all eight fundamen-

Figure 2: Spatial distribution of the most abundant value orientation groups in the 17 districts of Mannheim (source: own analysis)

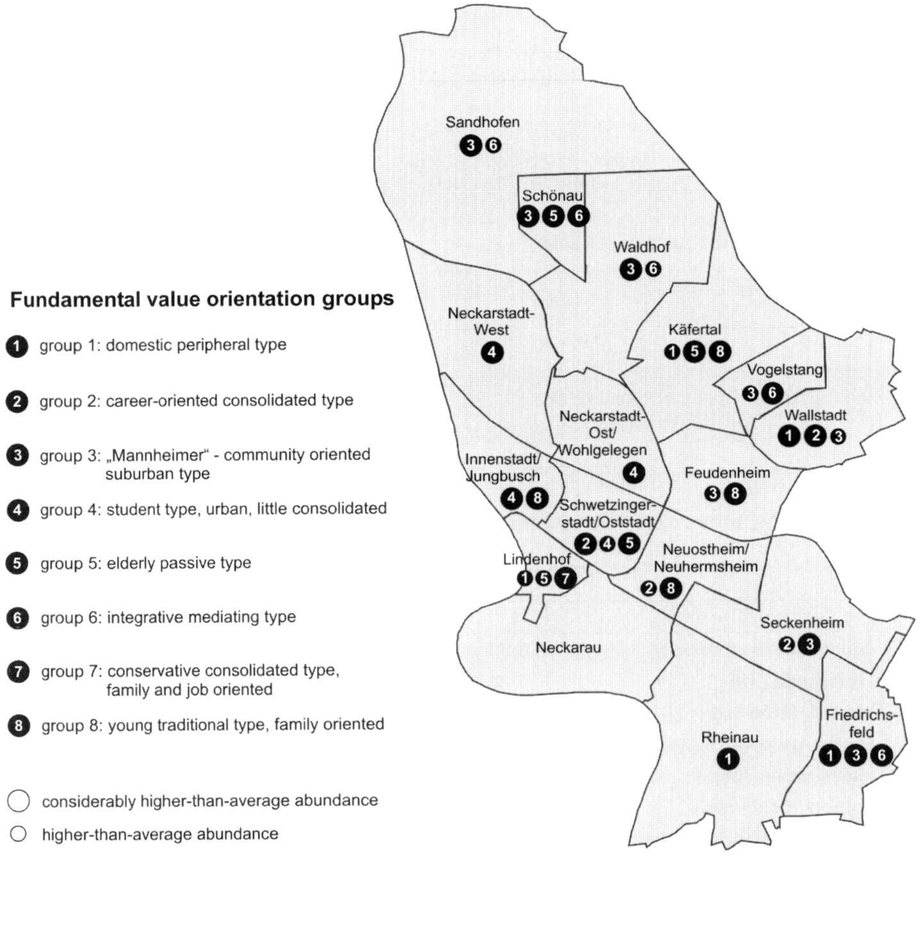

tal value orientation groups, which is displayed in figure 3. The eastern districts of Mannheim are characterised by a quite homogenous structure connected with high satisfaction values for all eight groups, especially in the district Feudenheim. In contrast, the northern and southern districts like

Figure 3: Average residential satisfaction of the value orientation groups

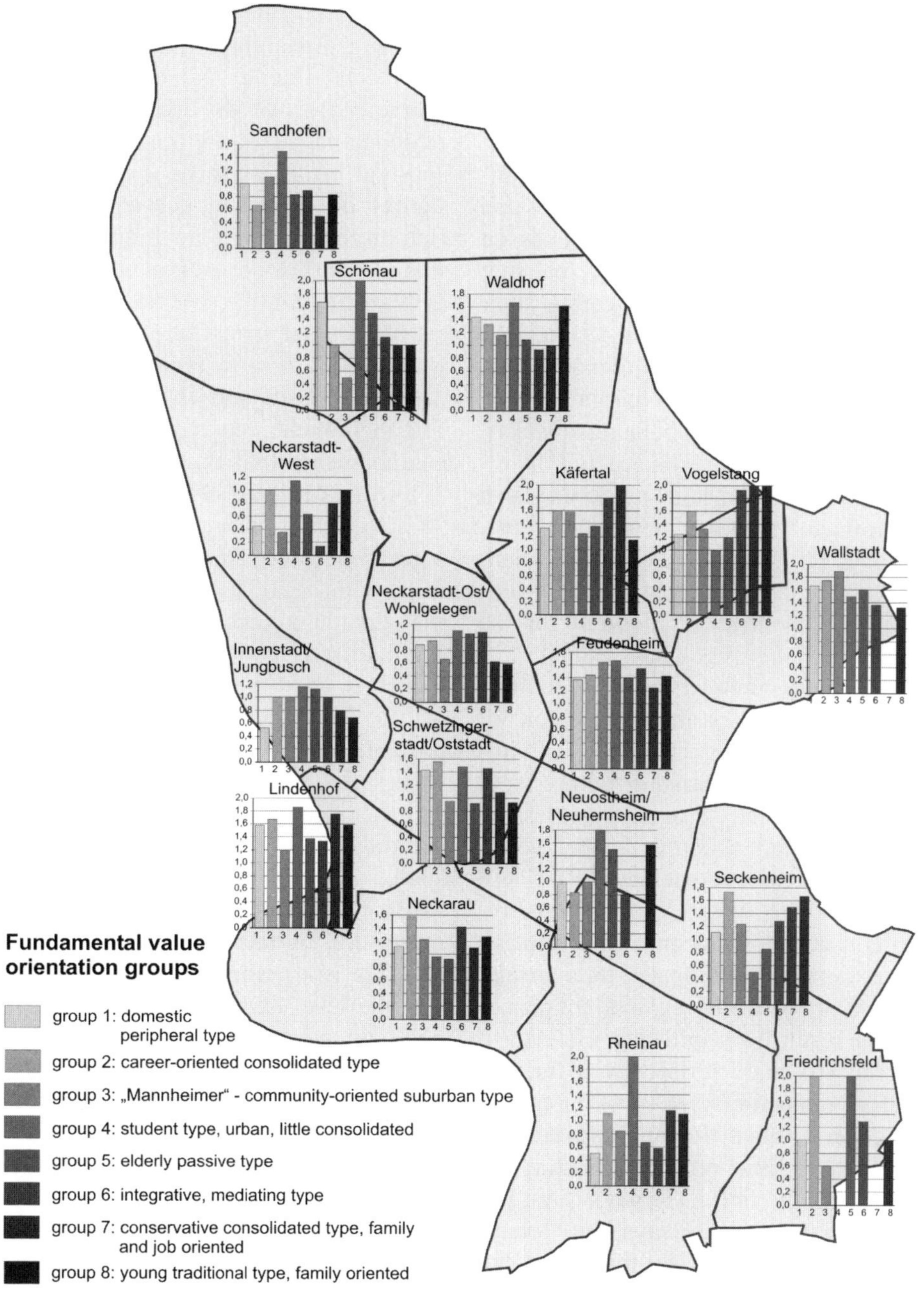

Sandhofen, Schönau, Rheinau Neuostheim/Neuhermsheim, Seckenheim or Friedrichsfeld show a more differentiated residential satisfaction. Neckarstadt-West is characterised by extremely heterogeneous values, with a very low value for group 6 but high values for the groups 2, 4, 7 and 8.

Spatial Econometrics

The residential satisfaction of the individuals belonging to the different value orientation groups mainly depends on their place of residence. For realising their requirements and meeting their preferences, each value orientation group might use the facilities and look and feel the social and physical characteristics of the neighbouring districts. Thus, the place of residence and the resulting residential satisfaction may also be influenced by the surrounding area. This effect is called "spatial dependency", and, consequently, spatial econometric models are required for the analysis "Who lives where and why?".

Spatial econometrics is a subfield of econometrics, which focuses on location, spatial interaction (spatial autocorrelation) and spatial structure (spatial heterogeneity): It uses regression models which account for the contiguity of regions (Anselin 1988). These spatial effects would lead to biased results if traditional econometric approaches were applied. Spatial autocorrelation describes interregional influences which result from spill-over effects and which are characterised by a functional relationship between two regions. In general, two different interpretations for such a functional relationship can be found: the first is Tobler's "first law of geography" which says that "Everything is related to everything else, but near things are more related than distant things." (Tobler 1970). The scond explains spatial autocorrelation as a consequence of the area unit problem, i.e. as an artefact caused by incorrectly delineated regions (Openshaw/Taylor 1979). Spatial heterogeneity describes a structural instability of the regression parameters due to the heterogeneous characteristics of the spatial units. In an econometric setting this causes non-constant error variances (heteroscedasticity) or biased model coefficients. Heteroscedasticity is an indication for spatial clusters or regimes (Anselin 1999).

The spatial structure of Mannheim is modelled using a binary contiguity matrix W (Cliff and Ord 1973; 1981). The matrix elements corresponding to neighbouring districts are denoted by "1", while all other elements are set to "0". By definition, a district cannot be a neighbour of itself, so the diagonal is filled with Nulls. For better interpretation, the spatial contiguity matrix is row standardised (spatial weights matrix).

Local Indicators of Spatial Association

The Local Indicators of Spatial Association (LISA) are used for detecting spatial dependencies on a local level as they separately measure contiguous influences for each district. The use of local statistics is motivated by the fact that the level of spatial dependence is not homogenous but differs over the regions due to spatial heterogeneity. This is called "spatial non-stationarity" (Kosfeld/Eckey/Türck 2007). LISA helps to detect spatial concentrations and thus spatial clusters of above and below average values of a variable.

The Local Moran coefficient l_i (Anselin 1995, p. 98 ff.) for district i uses the row standardised spatial weights matrix to detect local spatial dependence. The numerator in equation (1) is the product of the deviation of the "residential satisfaction" y_i from its average value and its arithmetic mean over the contiguous districts. For normalisation the term is divided by the average squared deviation of the residential satisfaction of the n observations:

$$l_i = \frac{(y_i - \bar{y})\sum_{j=1}^{n} w_{ij}(y_j - \bar{y})}{\sum_{j=1}^{n}(y_j - \bar{y})^2 / n} \qquad (1)$$

The Local Moran coefficient l_i shows a high positive value if district i and its neighbours all have above or below average values of Y (high-high or low-low cluster), while high negative values indicate dissimilar values of district i and its neighbours (high-low or low-high cluster). Values close to zero indicate that district i shows a value y_i which is close to the average of its neighbours or of Mannheim (Eckey et al. 2007).

The significant Local Moran clusters for all eight fundamental value orientation groups for the 17 districts of Mannheim are shown in Figure 4. Above average values of the residential satisfaction of most of the value orientation groups are concentrated as high-high-clusters in the eastern districts Käfertal, Vogelstang, Wallstadt and Feudenheim. The western districts are characterised by significant low-low clusters. Neckarstadt-West can be considered as a cold spot for groups 1, 5, 6 and 7 because the residential satisfaction there and in the adjacent districts is below the average of Mannheim.

Vogelstang is a low-high cluster for group 1. Obviously the members of this group cannot meet all their requirements with respect to their residential environment inside the district, in contrary to the adjacent districts Käfertal and Feudenheim-Vogelstang is chosen as residence by necessity due to personal, household or market dependent constraints, e.g. low-income households depending on affordable housing or international migrant households who are affected by constrained access to parts of the housing market. Opposite circumstances are found in Lindenhof: the members of group 1 are able to realise their requirements here and deliberately chose this district, which consequently forms the core of a high-low cluster.

Econometric Models

Spatial econometric models are used to identify which dimensions of the fundamental value orientation determine the residential satisfaction of each value orientation group.

A widely used approach to describe a functional relationship (spatial autocorrelation) between several spatial units is to specify a spatial stochastic process for a random variable at a given location (Anselin 1988: 17 ff). Similar to time series analysis, a lag operator (the elements of the spatial weights matrix) is used, but not as a shift along the time axis but as a shift in space.

Let us consider the linear regression model

$$Y = X\beta + \varepsilon \qquad (2)$$

where Y is the vector of the 17 observed values y_i of the residential satisfaction in the districts (dependent variable), X is the matrix of the seven extracted value dimensions (explanatory variables)

Figure 4: Local Moran coefficients for the 17 districts of Mannheim per value orientation group (source: own analysis)

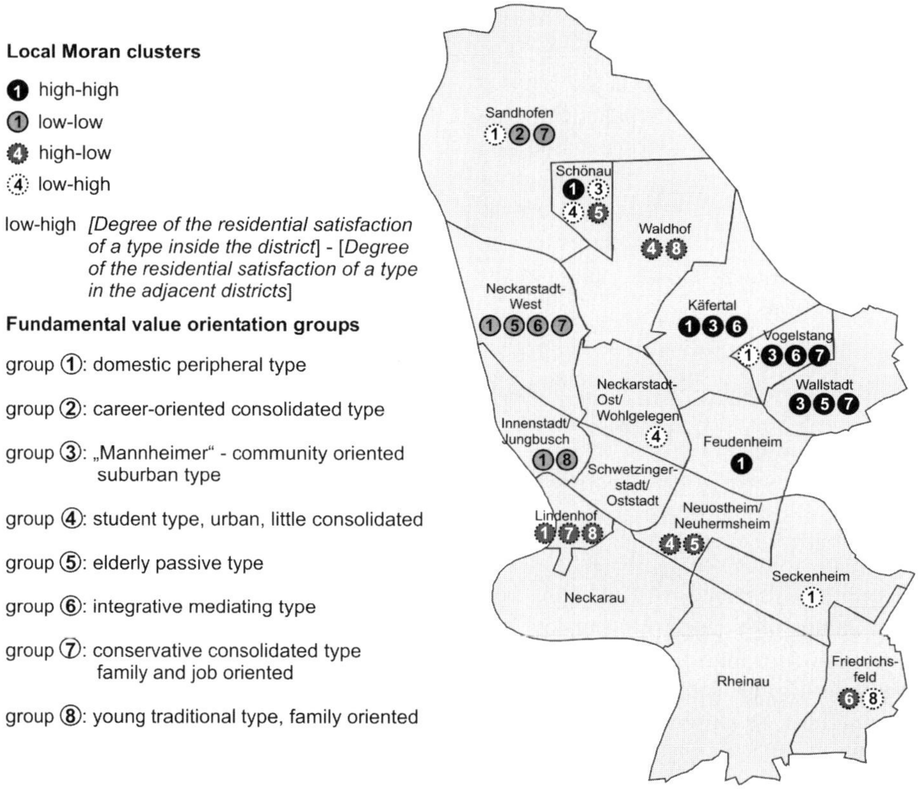

for all 17 districts, β is the associated 7 by 1 vector of regression coefficients and ε is a vector of random errors.

In a setting of spatial dependence the spatial structure was modelled by using the spatial weights matrix W and including a spatial stochastic process into the regression. If there is a significant correlation, it would be possible to make the decision which one of the two standard spatial econometric approaches has to be applied by using the series of LM tests. These two basic models are the spatial lag (equation 3) and the spatial error model (equation 4). In both cases, a spatially lagged variable is included into the (maximum likelihood) regression specification. It is inserted into the first model in form of an additional exogenous variable; in the second as part of the disturbance term (Eckey et al. 2006). The spatial lag model is specified as:

$$Y = \rho WY + X\beta + \varepsilon. \quad (3)$$

This model is appropriate if the focus of interest is the assessment of the existence and strength of spatial interaction. This is interpreted as substantive spatial dependence in the sense of being directly related to a spatial model. The spatial lag WY can be interpreted as the weighted average of the residential satisfaction of contiguous districts, while ρ is the spatial autoregressive coefficient that quantifies the relationship between neighbours and their residential satisfaction.

The spatial error model is the following:

$$Y = X\beta + \lambda W \varepsilon + u, \qquad (4)$$

λ is the spatial autoregressive parameter and u is a vector of i. i. d. (independent and identically distributed) errors. The spatial lag for the errors can be explained by random shocks, omitted variables or by the area unit problem (Eckey et al.2006).

Results of the Econometric models

Econometric modelling including spatial dependency analysis was performed for each of the fundamental value orientation groups using both the spatial lag and the spatial error model. Both models are indicated by the usual specification tests: the null hypothesis of the Wald Test, which assumes that the model has no explanatory power, is clearly rejected and the Akaike Information Criterion (AIC), which compares the models to a specification without spatial focus (linear model), concludes that the chosen specifications are superior compared to the linear model for each group. The choice between the spatial error model and the spatial lag model was taken on the basis of a robust LM test.

As a consequence of the low number of observations (17 spatial units) the statistical significance of the results is not high enough for all fundamental value orientation groups. In case of the spatial lag model (eq. 3), which turned out to be superior for groups 1, 3, and 6, group 1 shows highly significant values for β and ρ (table 2). The positive significant value of ρ (p-value: 0.02) clearly demonstrates a spill-over effect between neighbouring spatial units, here districts, i.e., a high value of the residential satisfaction y_i of district i has a positive effect on the satisfaction value of the direct neighbouring units.

Furthermore, for value orientation of group 5 significant model parameters could be obtained for the spatial error model (p-value: 0.06) (table 3), which also turned out to be superior for groups 2, 4, 5, 7, and 8. For group 5, negative correlation was found for errors in neighbouring districts (λ = -0.914) which might be due to the area unit problem (ch. 4) or the low number of districts analysed.

For both exemplified value orientation groups, several regression coefficients corresponding to different fundamental value orientations are significant to the five per cent level of confidence (marked as bold in tables 2 and 3) and thus can be interpreted as determinants of the observed residential satisfaction values. These relations are not necessarily causal in the sense that the prevailing values directly influence residential satisfaction, as they could also just reflect spatial correlation in the value orientations.

Nevertheless, the results allow a more profound characterisation of the value orientation groups and their impact on residential preferences and satisfaction. For both groups, a positive influence of the socio-economic capacity of the other residents on their residential satisfaction is clearly visible, which is obviously independent from the own economic situation (cp. table 1): while group 1 possesses a higher-than-average socio-economic capacity, group 5 is provided with rather low socio-economic capacity (cp. table 1), but for both groups the residential satisfaction is increasing significantly with an increasing socioeconomic capacity in the district (β=0.3737 for group 1 (table 2) and β=0.4543 for group 5 (table 3). This tendency is also observed for the other value orientation groups which show a significant value of this coefficient (groups 2, 7 and 8).

The dimensions "conception of society" and "tolerance" provide information on how the inhabitants subsumed to group 5 are orientated with respect to social life and gender role (table 3). Therefore, the preference for neighbours with the same traditional orientation (conception of society) can be easily understood, the same holds for their affinity to neighbours with an urban orientation and a lower self-confidence level.

Instead, the apparent attractiveness of traditional orientations and lower self-confidence to group 1 (positive sign in table 2), which is characterised as modern and rather self-confident (cp. table 1), can only be understood as the appeal of difference, which might be predominant for people with a higher level of education, self-confidence and creativity (West 2007). Unfortunately, the actual data do not allow confirming this presumption.

Table 2: Group 1 (Domestic Peripheral Type): Spatial Lag Model

Dimension	β	Std. Error	Z value	Pr(>|z|)
conception of society (+)/(-)	0.173547	0.064420	2.6940	0.007060
tolerance (+)/(-)	-0.072648	0.048827	-1.4879	0.136785
self-determined (+)/(-)	-0.022459	0.076329	-0.2942	0.768577
self-confident (+)/(-)	-0.386222	0.068345	-5.6510	1.595e-08
suburban/urban (+)/(-)	0.213709	0.073324	2.9146	0.003562
global/local (+)/(-)	0.163449	0.049676	3.2903	0.001001
socio-economic capacity(+)/(-)	0.373775	0.079399	4.7075	2.507e-06

Rho: 0.49482 — LR test value: 4.912 — p-value: 0.026671
Wald statistic: 6.6394 — p-value: 0.0099746
AIC: -2.0269 — (AIC for lm: 0.88503)

Table 3: Group 5 (Elderly Passive Type): Spatial Error Model

Dimension	β	Std. Error	z value	Pr(>|z|)
conception of society (+)/(-)	0.44946786	0.08098511	5.5500	2.857e-08
tolerance (+)/(-)	0.11061326	0.06025368	1.8358	0.066388
self-determined (+)/(-)	-0.00090889	0.11152955	-0.0081	0.993498
self-confident (+)/(-)	-0.17510810	0.07706205	-2.2723	0.023068
suburban/urban (+)/(-)	-0.25595517	0.09507861	-2.6920	0.007102
global/local (+)/(-)	-0.06960859	0.05191952	-1.3407	0.180017
socio-economic capacity (+)/(-)	0.45430189	0.10072367	4.5104	6.471e-06

Lambda: -0.91403 — LR test value: 3.535 — p-value: 0.060085
Wald statistic: 7.6261 — p-value: 0.005753
AIC: 2.1375 — (AIC for lm: 3.6725)

Summary and Conclusion

Based on a survey on migration in Mannheim, Germany, a study on value orientations, choice of residence and residential satisfaction was performed. Besides general data on households, individuals and orientations, housing needs, moving motives and residential satisfaction was inquired. In a first step, seven fundamental value orientations were constructed using different variable subsets and exploratory factorial analysis which allowed for a characterisation of the inhabitants in the 17 districts of Mannheim. On the basis of these fundamental value orientations, the total of 2,508 surveyed inhabitants was subdivided into eight value orientation groups using cluster analysis.

In a second step and based on the value orientation groups constructed before, econometric modelling of the observed residential satisfaction as function of the fundamental value orientation of the districts was performed including spatial dependence between neighbouring districts. Local Moran coefficients indicate local dependence and were interpreted in terms of hot and cold spots (high-high, low-low or high-low and low-high clusters). Spatial econometric modelling was performed by using either the spatial lag or the spatial error model. As expected, significant model parameters could not be obtained for all value orientation groups, but for most of the dimensions in case of groups 1 and 5 ("domestic peripheral type" and "elderly passive type"). This is probably due to the low number of cases (districts) used for modelling. Nevertheless, the principle applicability of the method could be demonstrated and a deeper characterisation of the typical members of the value orientation groups, especially for group 1 and 5, could be achieved.

In conclusion, the explorative approach based on fundamental value orientations of the residents combined with a confirmative spatial econometric modelling turned out to be a promising tool for measuring residential needs and trends and to find out "Who lives where and why?". The method is largely model-free, i.e., no a priori assumptions on actual trends or lifestyle groups are necessary. The practical applicability of the new method crucially depends on a sufficient number of observations (number of districts and total number of surveyed inhabitants) which was not fully provided in the presented case.

References

Anselin, L. (1988): Spatial Econometrics: Methods and Models. Kluwer Academic Publishers, Dordrecht, The Netherlands.

Anselin, L. (1999): Spatial Econometrics. Available at: http://www.csiss.org/learning_resources/content/papers/baltchap.pdf (21.12.09)

Anselin, L., Bera, A., Florax, R. & M. Yoon (1996): Simple diagnostic tests for spatial dependence. In: Regional Science and Urban Economics 26, pp. 77–104.

Bourdieu, P. (1982): Die feinen Unterschiede. Kritik der gesellschaftlichen Urteilskraft. Suhrkamp, Frankfurt a. Main.

Cliff, A. & J. K. Ord (1973): Spatial Autocorrelation. London.

Cliff, A. & J. K. Ord (1981): Spatial processes: Models and applications. London.

Eckey, H.-F., Kosfeld, R. & M. Türck (2006): Räumliche Ökonometrie. WiSt Wirtschaftswissenschaftliches Studium, Heft 10, pp. 548-554.

Gans, P., Schmitz-Veltin, A. & C. West (2009): Wohnstandortentscheidungen von Haushalten am Beispiel Mannheim.

In: Raumforschung und Raumordnung 68, Heft 1, S. 49-59..

Kosfeld, R., Eckey, H.-F. and M. Türck (2007): LISA (Local Indicators of Spatial Association). WiSt Wirtschaftswissenschaftliches Studium, Heft 3, pp. 157-162.

Meng, R., Schmitz-Veltin, A. & C. West (2008): Wohnen in der Stadt? – Wohnwünsche intraurban wachsender Haushalte und potenzieller Reurbanisierer am Beispiel der Stadt Mannheim. In: Maretzke, S. (ed.): Städte im demographischen Wandel – Wesentliche Strukturen und Trends des demographischen Wandels in den Städten Deutschlands, Materialien zur Bevölkerungswissenschaft, Heft 125. Wiesbaden, pp. 103-112.

Openshaw, S. & P. J. Taylor (1979): A Million or So Correlation Coefficients: Three Experiments on the Modifiable Areal Unit Problem. In: Statistical Applications in the Spatial Sciences, London: Pion, pp. 127-144.

Tobler, W. R. (1970): A computer movie simulating urban growth in the Detroit region. Economic Geography vol. 46, no. 2, pp. 234–240.

West, C. (2007): Dimensionen des individuellen und gesellschaftlichen Wandels: Die grundlegenden Orientierungen „geschlossen-offen“ und „konkret-abstrakt“ als Determinanten der Stadtentwicklung in Barcelona (1986-2005). Dissertation Universität Mannheim.

West, C., Gans, P. & A. Schmitz-Veltin (2008): Cities in space and city as place – Rethinking reurbanisation: urban, semi-urban and suburban orientations and their impact on the choice of residence. In: Zeitschrift für Bevölkerungswissenschaften, Band 33, Heft 3-4, pp. 381-408.

List of contributors

Dipl. Volksw. Philipp **Deschermeier**
University of Mannheim, Department of Economics
email: philipp.deschermeier@uni-mannheim.de

Univ.-Prof. Dr. Rudolf **Giffinger**
Vienna University of Technology, Centre of Regional Science
email: giffinger@srf.tuwien.ac.at

Dipl.-Geogr. Roland **Götzke**
University of Bonn, Department of Geography
email: r.goetzke@geographie.uni-bonn.de

Dr.-Ing. Alexandra **Hill**
Technical University Dortmund, Institute of Spatial Planning
email: alexandra.hill@tu-dortmund.de

Dr. Michael **Judex**
University of Bonn, Department of Geography
email: m.judex@geographie.uni-bonn.de

Univ.-Prof. Dr. Andreas **Koch**
University of Salzburg, Department of Geography and Geology,
Research Group Social Geography
email: andreas.koch@sbg.ac.at

Univ.-Prof. Dr. Reinhard **König**
Bauhaus-University Weimar, Computer Science in Architecture
email: reinhard.koenig@archit.uni-weimar.de

Dr.-Ing. Christian **Lindner**
Technical University Dortmund, Institute of Spatial Planning
email: christian.lindner@tu-dortmund.de

Ass.-Prof. Dr. Peter **Mandl**
University of Klagenfurt, Department of Geography and Regional Science
email: peter.mandl@uni-klu.ac.at

Dipl. Ing. Roman **Seidl**
Vienna University of Technology, Centre of Regional Science
email: seidl@srf.tuwien.ac.at

Christina West, **M.A.**
University of Mannheim, Department of Economics
email: west@rumms.uni-mannheim.de